The
Tenth
Dimension

The Tenth Dimension

An Informal History of High Energy Physics

Jeremy Bernstein

McGraw-Hill Publishing Company
New York   St. Louis   San Francisco   Auckland
Bogotá   Hamburg   London   Madrid   Mexico
Milan   Montreal   New Delhi
Paris   São Paulo   Singapore
Sydney   Tokyo   Toronto

Library of Congress Cataloging-in-Publication Data

Bernstein, Jeremy,
  The tenth dimension.

  Includes index.
  1. Particles (Nuclear physics)--History. I. Title.
II. Title: 10th dimension.
QC793.16.B47 1989   539.7′21′09   88-36407
ISBN 0-07-005017-1

1234567890   DOC/DOC   89432109

ISBN 0-07-005017-1

The editors for this book were Jennifer Mitchell and Galen H. Fleck
and the production supervisor was Richard A. Ausburn. This book
was set in Baskerville. It was composed by the McGraw-Hill
Publishing Company Professional & Reference Division composition
unit.

Printed and bound by R. R. Donnelley & Sons Company.

*For more information about other McGraw-Hill materials
call 1-800-2-MCGRAW in the United States. In other
countries, call your nearest McGraw-Hill office.*

Contents

Preface

It was just twenty years ago when, for the then Atomic Energy Commission, I wrote a small, semipopular booklet which I called *The Elusive Neutrino*. Much to my surprise, it went through several printings and, in the end, over a hundred thousand copies of it were distributed. In it I tried to describe both the history and the physics of the neutrino. I had in mind a reader who, while having no specialized scientific knowledge, had sufficient curiosity to want to learn about an important and fascinating scientific subject in a way that was neither misleading nor condescending. The fact that the booklet was so widely read persuaded me then that a substantial body of such readers exists.

In this book I have tried to do something much more ambitious for the same or a similar set of readers. I have tried to describe, at least in outline, both the history and the science of the entire field of elementary-particle physics and cosmology. (As the reader of this book will learn, these two apparently disjoint subjects now have a frontier in common.) These are fields in which I have been active since my graduate school days in the 1950s. Much of the history I describe I was witness to; and as I wrote about many of the people involved, I could see and hear them in my mind's eye and ear. In this book too I have tried to be neither misleading nor condescending. I have done my best to simplify, as much as I am able, some extremely difficult scientific material. I believe a reader who persists will be rewarded with a new level of scientific understanding.

I am grateful to many people for their help with this book. To begin with, I would like to thank Bill Hess of the Department of Energy and Nick Samios of the Brookhaven National Laboratory for suggesting that I write this book in the first place. The Department of Energy provided the support that enabled me to take a semester off from teaching so that I could do the research and writing. Herb Kinney of Brookhaven helped me with the administrative details of that arrangement, and I thank him.

Many of my colleagues read and criticized drafts of this book. Here is a list: C. B. Bratton, L. Brown, M. Gell-Mann, R. Jackiw, A. Jones, M. Turner, and J. C. Van der Velde. I am especially grateful to Gerald Feinberg and Maurice Goldhaber, who read the entire manuscript in detail and saved me from all sorts of folly.

Jeremy Bernstein

Prologue: The Large and the Small

On the evening of February 23, 1987, Albert Jones of Nelson, New Zealand, as was his custom, went out after dinner to the driveway near his home. In the driveway he had set up a large astronomical telescope of his own construction. Mr. Jones, who was then 66, had retired not long before from work in the repair division of an automobile company in Nelson. But since 1944, he had spent much of his free time, as an amateur astronomer, studying variable stars—stars whose brightness varies in the course of time. It is estimated that, by 1987, he had made about 350 000 variable-star observations. He has been called the world's greatest amateur observational astronomer, and in the summer of 1987 he was awarded the Order of the British Empire for his work. On February 23 he began his star scan by examining a relatively nearby galaxy known as the Large Magellanic Cloud, which, at about 9:30 local time—the time he was looking at it—was quite high in the sky. He noticed nothing unusual.

In view of what was about to happen, it is instructive to quantify Mr. Jones's nonobservation. His home-constructed telescope is what is known as a 12.5-inch Newtonian reflector. This means that in essence it is a long tube into which light enters and is then collected and focused by a mirror, in this case 12.5 inches in diameter. The mirror is one of the few things designed for a telescope which Mr. Jones actually bought. The light is then reflected out the side of the telescope by a second mirror and can be examined through an eyepiece. In Mr. Jones's case, the eyepiece is a World War II bombsight.

The telescope is powerful enough to detect stars up to what astrono-

mers call the 13.5 magnitude. Astronomers have traditionally utilized a
magnitude scale which, perversely, is so adjusted that the larger the
number the *weaker* the apparent brightness. Negative numbers are assigned to strong sources. Thus, on a logarithmic scale, the sun has an
apparent magnitude of −26.6; the Moon, −12.6; and Venus, at its
brightest, −4.4. Mr. Jones could have detected any peculiar star with a
magnitude less than 7.5, and he saw none in the Large Magellanic
Cloud that evening.

The next morning, as he often did, he set his alarm clock for a pre-
dawn hour so that he could continue his observing. By that time the
Large Magellanic Cloud was setting and, because he had already studied it that evening, he passed it by. That, as it turned out, was a pity.
Someone who had continued to watch for an hour or so after Mr. Jones
had turned his attention elsewhere would have seen the most spectacular astronomical event to occur in our vicinity in the past 383 years. The
unprepossessing blue star Sk-69° 202, so designated in a catalog by
Nicholas Sanduleak of the Case Western Reserve University, had suddenly become a supernova—SN 1987 A—so bright that it would have
been visible with the naked eye. Many supernovas are observed each
year, but this was the first one in nearly four centuries seen to explode
in our own or a close-by galaxy.

The following morning Mr. Jones was back at his telescope. Now, to
sort out the sequence of events, it is useful to describe things in what
astronomers call Universal time (UT). (It is also known as Greenwich
mean time.) The advantage of doing so is that the effect of the longi-
tude gets subtracted out and one can see, at a glance, which event is earlier and which is later. In UT units, Mr. Jones's initial nonobservation
was recorded at 9:22 on February 23. It was now 8:52 UT the following
evening. When Mr. Jones pointed his telescope at the Large Magellanic
Cloud, the supernova—in his words—"popped out" at him. By that time
it was at + 4.4 magnitude, which is to say it was as bright as a star visible
to the naked eye. Mr. Jones at once called Frank Bateson in nearby
Touranga. Mr. Bateson has been, for over sixty years, the director of
the Variable Star Section of the Royal Astronomical Society of New
Zealand. Mr. Jones said to Mr. Bateson, "Frank, there is a star in the
Large Magellanic Cloud where there was no star before." Mr. Bateson
would have telexed this remarkable news to the International Astronomical Union in the United States but, as it happens, no commercial
telexes can be sent from New Zealand at night. He did, however, get
word immediately, by telephone, to astronomers in Australia.

In the meanwhile, a Canadian observer who is a professional astronomer, Ian Shelton of the University of Toronto, working at the Los
Campanas station in Chile, happened to develop a photographic plate

at 5:31 UT, some three and a half hours earlier. The photographic plate was of the Large Magellanic Cloud, and in it was an image of what appeared to be a very bright star. At first Shelton thought he was looking at a flaw in the plate. He went outside to look, and, sure enough, the new supernova was visible to the naked eye. Thus Shelton became the second person to discover SN 1987 A, with Mr. Jones a close third. The first person to actually see it, an hour and a half earlier, was another Las Campanas observer named Oscar Duhelda, who did not tell anyone until Shelton brought the matter up. Because he did *not* see the supernova the night before, Mr. Jones was able to narrow down the time of the explosion. But as it happened, unknown to anyone at the time, remnants of the supernova first made their appearance here on earth at 7:35:41 UT on February 23, nearly two hours prior to Mr. Jones's non-observation. The remnants were in the form of ghostly, elusive particles called neutrinos.

At the Fairport mine in Painesville, Ohio, some 20 miles from Cleveland, miners and other employees of the Morton Salt Company— "When it rains, it pours"—tend to wear patches on their overalls and miners' helmets which read "Think Snow!" As it happens, that is not because these people have an affinity to skiing. Rather, it is because they mine rock salt, which is sprinkled on icy roads—when it pours they reign. The Fairport mine is very much a working commercial mine. Visiting it is a serious matter. The increasingly uneasy visitor is given a safety lecture which includes an interlude on using a rather heavy catalytic converter to be worn at all times, on a thick leather belt, when the visitor is underground. In case of a mine fire, the converter transforms carbon monoxide, which is poisonous, into carbon dioxide, which isn't. It also gets very hot when it is working, and visitors are told that, despite the heat, they had better keep converters in their mouths if they are going to survive. Each visitor is then given a pair of hightop shoes with iron toes as a precaution against tripping over lumps of salt, a pair of blue coveralls, and a hard plastic helmet with a miner's lamp attached to it. He or she is then ready to go underground.

The mine shaft is 2000 feet deep—nearly two Empire State Buildings. There are two elevators that connect the bottom to the surface: a "freight" elevator—a sort of bucket—that brings up the salt and can be used, in emergencies, for people, and a passenger elevator which is big enough to hold 10 miners on each of its decks—about 40 at a time work in the mine—and, if necessary, various kinds of material. It is a comfortable, if austere, steel cage. The ride down takes about four minutes and feels, to a newcomer, like a small eternity. The temperature increases with the increase in depth. The walls of the working face of the mine are at a temperature of about 85°F. There is a good deal of humidity at the shaft bottom. Murky tunnels seem to lead in all directions. The visitor passes a huge, noisy fan and is led down a dark tunnel, with a miner's lamp illuminating the way. Care must be taken not to slip on the humid salt. After walking for what seems like another small eternity, the visitor is confronted by what looks like an impassable steel wall. There are warning signs about electrical equipment, but with a proper key the iron barrier can be opened and passed. At first, the visitor is startled by the bright sodium vapor lights and the air-conditioning. He or she is in another world, the world of a high-energy physics laboratory.

This laboratory is devoted to a single enterprise: the care and feeding of an entity known as the IMB detector. The initials "IMB" stand for Irvine, Michigan, and Brookhaven: the University of California at Irvine, the University of Michigan at Ann Arbor, and the Brookhaven National Laboratory at Upton, Long Island, New York. Those three institutions built the detector; they now maintain and run it. What is the detector? In November 1979, the work on the presently occupied chamber—which is separate from the rest of the mining operation, although close enough that blasting can be felt in it—was started, and the excavation was completed August 26, 1980.

On entering the chamber, the visitor passes through a large room filled with what looks like, and is, a very complex water filtration system, the reason for which will become evident shortly. In the next room there is a kind of laboratory bench area, with, among other things, a microwave oven suitable for making coffee. In the next area there is a maze of computer equipment, and, finally, the visitor arrives at the heart and soul of the detector: a gigantic pool of transparent and highly purified water.

The tank is rectangular in shape; its width is 60 feet, its depth 65 feet, and its length 80 feet. It holds a total of 8000 metric tons of water. The water is so clear that when the divers from the University of Michigan, who service the tank about every three weeks, first went into it, they got vertigo. It was like falling through thin air. On looking into this water, one sees an array of 2048 photomultiplier tubes. These are very large— 8 inch-diameter—hemispherical tubes recently developed for this kind of detection of weak light signals. Each one costs about $1000. A couple of them probably cost about as much as Mr. Jones's entire telescope.

As the reader will have gathered from the name, the purpose of the photomultiplier tubes is to detect light. Therein lies a tale. The light— more generally, electromagnetic radiation—to be detected is emitted by rapidly moving electrons. If the electrons are moving more rapidly than the effective speed of light in water (nothing can move more rapidly than the speed of light in a vacuum, but a material medium slows the light up), they can emit what is known as Cerenkov radiation. (It is so called after the Russian scientist Pavel Aleksejevic Cerenkov, who shared the 1958 Nobel prize in physics for proposing it.) This radiation is emitted in a narrow cone — something like a sonic boom — with its axis in the direction of the electron's motion, a detail that is crucial to using it in a detector. By detecting the Cerenkov radiation, it is possible to reconstruct the trajectory of the particle that emitted it, in this case an electron.

The electrons that the detector was built to detect had, in the first instance, nothing to do with supernovae. In the middle 1970s several theoretical physicists proposed theories that unified previously disparate parts of elementary-particle physics — of this, more later. An inevitable consequence of the unification program, it appeared, was that matter had to be somewhat unstable. The proton, which had usually been taken as the stable building block of matter, would, it was thought, actually decay but with a very long lifetime — some 10^{31} years. Because that is much longer than the age of the universe, which is about 10^{10} years, not very many protons in an average sample of matter were predicted to decay in, say, a year.

The most prominent of the theories predicted a proton decay mode in which a very energetic positive electron would be emitted. Hence, one needed a detector that could detect a very rare event involving the emission of a very energetic electron by a decaying proton, ergo the tank. The 8000 tons of water in the tank contain about 2×10^{33} protons. If each proton lived even as long as 10^{31} years, in such a mass of protons there could still be about 200 decays a year. That would be an event every couple of days, enough to work with. One may well ask — having seen the reason for it — why build the tank in a mine chamber 2000 feet below the surface of the earth, a mine chamber which, incidentally, cost $200 000 to dig? (The physicists got a very good price because the Morton Salt people wanted to try out a new piece of digging equipment.) The reason is what the physicists call background. The earth's atmosphere is constantly bombarded by cosmic rays, and they can produce Cerenkov radiation — which can look just like what would be expected from proton decay. The mine shaft is located under 2000 feet of absorber, which screens out most of the background. The odd neutrino does get through, but it only produces an event that mocks up what is being looked for every two or three years.

The proposal to dig the tunnel was funded in 1979, and the digging was completed in the summer of 1980. The tank was filled completely with ultrapure water by July 30, 1982, and the experimenters began to wait for proton decay events. They are still waiting. In the words of Maurice Goldhaber, one of the senior experimenters from Brookhaven,

"We had some candidates, but they weren't elected." At this writing, the conclusion is that the proton must live at least 3.5×10^{32} years, a disappointingly long time for some of the theorists who championed simple models of unification that predicted shorter times.

In order to understand what happened on February 23, we must return, once again, to the tank. Occasionally, very occasionally, high-energy neutrinos that course through it, because of cosmic-ray phenomena, hit one of the protons in the water molecules that make up the content of the tank and produce a positron, a positive electron. Now, the velocity of light in water is about three-quarters of the velocity of light in vacuum, which is, in turn, about 300 million meters a second. If the positron emerges from its interaction with the neutrino with enough energy to be moving faster than the velocity of light in water, which is about 70 percent of its speed in vacuum, it will Cerenkov-radiate. The greater its energy the more intense will be the radiation; and if it has enough energy, it will trigger enough of the 2048 photomultiplier tubes to be detectable. The signals from those tubes are carried over wires into the neighboring room, where they are fed into computers which make preliminary analyses of them. The process takes place automatically and goes on night and day as long as the detector is in service, whether or not anyone is actually down in the mine to observe it.

It was going on Sunday night, Cleveland time, and during the early hours of the morning of February 23, which happened to be a Monday. That is why no one was there to notice that a power supply had tripped off, shutting down a quarter of the phototubes. In the meantime, it was late in the afternoon in Kamioka, Japan. In Kamioka there is another detector—the Kamiokanda II, a modified version of the IMB—in a zinc mine. It has a tank with 3000 tons of water but with phototubes that cover a larger fraction of the surface, which means that electrons, and hence neutrinos, of lower energy can be detected. At precisely 7:35:35 UT—within an error of 50 milliseconds—both detectors recorded bursts of energetic neutrinos. The Kamiokanda II detector recorded 11 neutrinos in 12.439 seconds; the IMB detector recorded 8 in 5.59 seconds.

The reader should understand that all of this was realized only several days after the fact. That it was realized at all is a tribute to the interplay between theory and experiment. In the first place, over the past two decades a picture of how a supernova develops has evolved. For a star to become a supernova it must be more massive than what is known as the Chandrasekhar limit. (The limit is named after the Indian-born American astrophysicist Subrahmanyan Chandrasekhar, who won the 1983 Nobel prize in physics for this and other fundamental contributions to theoretical astronomy.) The "classical" Chandrasekhar limit is

1.456 times the mass of the sun, but subtle corrections can make it as low as about 1.2 times or as high as about 2 times that mass. Fortunately for us, or our descendants, it is always larger than the mass of the sun, because a star that has a mass that exceeds it is at the end of its evolution and is doomed to collapse and, in the case of a supernova, to explode.

Prior to its explosion, such a star consists of a dense iron core—iron because the fusion reactions that power a star cannot produce elements heavier than iron—surrounded by onion skin layers of lighter elements; typically, moving outward, they are silicon, oxygen, carbon, helium, and hydrogen. Once the iron core becomes sufficiently massive and sufficiently hot, it simply collapses in a fraction of a second—milliseconds. At first the outer layers do not know that the inner core has collapsed. Then they are hit by a stupendous shock wave. The final fate of the inner core is to become a neutron star, a crystalline mass of neutrons, denser than an atomic nucleus, weighing about the mass of the sun and only about ten kilometers across.

To accomplish that, and still conserve energy, a mind-boggling amount of energy must be released: about 3×10^{53} ergs in about a second. To have some perspective on that number, note that the sun—no mean radiator—emits energy at a rate of "only" 3.9×10^{33} ergs per second. The emission rate in supernova formation is therefore equivalent to 10^{20} suns. The general belief of the scientists who study such matters is that at least 99 percent of the energy is emitted as neutrinos, not in the initial explosion, but as a consequence of processes triggered by the resultant shock wave. The supernova turns into a sort of neutrino star. These neutrinos, in the case of SN-1987 A, took about 163 000 years to reach our planet. There they arrived on February 23, some three hours before the visible light also produced by the shock wave.

About, it is thought, 10^{58} neutrinos were emitted in the explosion. About 30 million billion of them passed through an IMB detector in 6 seconds, and only 8 of them made suitable collisions and were detected—after the fact. Nothing in the IMB circuitry as it was then—it has since been modified—would have instantaneously signaled to an experimenter that such a neutrino burst had taken place. In the future, a computer readout will say "possible neutrino burst detected at time...," if there is one. Both the IMB people and the physicists at the Kamiokande II were alerted to the possibility that neutrino events could be found retrospectively in their data by the accounts of the discovery of Duhelda, Shelton, and then Jones of the visible supernova. As it happens, we were lucky to have found those neutrinos at all. As mentioned, the IMB had a power surge which shut off a quarter of its phototubes and, as it turned out, the Kamiokanda II suffered a complete power outage on February 25, which made it impossible to make an absolute calibration of the time the neutrinos arrived there. It was then, in this serendipitous way, that the subject of extragalactic observational neutrino astronomy was founded.

1
The Weak and the Strong

Sunday newspapers used to have – perhaps still have – Sunday supplements which featured something called "What is wrong with this picture?" It was a sort of cartoon which, for example, might show an automobile. The reader who looked carefully noticed, say, that one of the door handles was on backwards. Perhaps the feature should have been called, "What is paradoxical about this picture?" since there is nothing actually "wrong" about having a door handle on backwards. In reviewing the Prologue we might well ask the same question. Let us reconsider some of the facts. Here are three:

1. About 30 million billion neutrinos went through the IMB detector in 6 seconds; only 8 appear to have interacted with the electrons in the water molecules.

2. The neutrinos arrived here about 3 hours *before* the light, although both were moving with about the same speed – after all, the difference in arrival times was only 3 hours in 163 000 years. Both originated in the same explosion.

3. No observer at the IMB detector could possibly *see* the Large Magellanic Cloud, which is visible only in the southern hemisphere. The earth blocks the light, yet the stream of neutrinos from the supernova passed through the earth as if it weren't there.

If we knew nothing else about neutrinos except those three facts, we would already know that there is something very different about the way neutrinos interact with matter and the way light does. It is obvious that, comparatively speaking, neutrinos hardly interact with matter at all. In that case, how were they ever discovered? Thereon hangs a tale.

The neutrino began its existence in the mind of Wolfgang Pauli. It would appear, from correspondence, that the idea occurred to Pauli during the week of November 26, 1930. Before explaining what led Pauli to the idea, it is worthwhile to say a few words about the man himself. Pauli was one of the most interesting of the twentieth century theoretical physicists. He was born in Vienna on April 25, 1900. His Viennese background may account for the fact that he knew the Strauss opera *Die Fledermaus* well enough to take one of the lines in it, "So young and already a prince," and transform it into a withering comment about another physicist, "So young and so already unknown." His father, Wolfgang, Senior, was a professor of colloid chemistry at the University of Vienna. Pauli was something of a prodigy, and by the age of 20 he was entrusted, by his university teacher, Arnold Sommerfeld in Munich, with the task of writing a review article about the theory of relativity. The article was so good that it can still be read with great profit by anyone seriously interested in the subject.

Pauli was not then, or at any other time, afraid to express his opinions about the work of others, no matter who they were. While still a student at a meeting at which Einstein had spoken, he began some comments with the remark, *"Was der Herr Einstein da gesagt hat, was gar nicht so dumm.* [What Mr. Einstein has said was not so dumb]." Nor had he any false modesty about his extraordinary abilities. When a young assistant informed him in 1952 that the distinguished Dutch physicist Hendrik Kramers had just died, Pauli thought for a minute and said, *"Der Kramers war gut...der Kramers war gut...aber Ich war besser.* [Kramers was good...Kramers was good...But I was better]." Be that as it may, for psychological reasons one can perhaps come to understand, he seemed to find his own neutrino hypothesis genuinely disturbing— almost frightening.

Pauli was faced with what he perceived to be a paradox in the class of radioactive decays in which electrons are emitted. These were, and still are, called beta-decays. In the same nomenclature, helium nuclei— which also are emitted in some decays—were given the name "alpha-particles." "Gamma-particles"—or "gamma-rays"—was the name given to energetic electromagnetic quanta. In a typical beta-decay the unstable, radioactive parent nucleus disintegrates into what appear at first sight to be only two daughter particles which carry electric charge. A decay in which two—and only two—particles are emitted is called a two-body decay. If the decaying particle is at rest when it decays, then, to conserve momentum and energy, the daughter particles must come out in opposite directions with fixed energies. Therefore, the beta-particle

must, in a two-body decay, come out with a single fixed energy characteristic of the particular decay. But that is precisely what did not happen. The beta-particles were observed to come out with an entire continuum, a "rainbow" of possible energies: an energy spectrum. That was
the paradox.

The matter was so desperate that, to deal with it, Niels Bohr proposed
giving up the conservation of energy. To Pauli that was anathematic.
Instead, Pauli invented—he made up—a new particle which he called
the neutron. This is not to be confused with what we now call the neutron, the object which, along with the proton, is taken to be a building
block with which atomic nuclei are made. *That* neutron was discovered
by James Chadwick in 1932. Pauli's neutron, as we shall see, was discovered, that is, directly observed, only in 1956. This "neutron," if it were
emitted in beta-decay along with the electron and daughter nucleus,
would, by taking off some of the energy, resolve the β-decay paradox
and make unnecessary the abandonment of the conservation of energy.
Pauli was also aware that it would resolve a more subtle paradox involving the conservation of angular momentum and a still more subtle paradox having to do with quantum mechanical statistical mechanics, all at
the expense of introducing a hypothetical particle no one had ever observed.

A measure of his reticence about all this can be gleaned from a letter
Pauli wrote on December 4, 1930, addressed to a group of colleagues
gathered in a meeting in Tübingen. We are fortunate to have even this
letter because, as it turned out, Pauli never published his "neutron" hypothesis. We have the letter only because of a glitch in his marital status.
He was between marriages, having been divorced that same week, and
had decided, therefore, to attend a student ball in Zurich. He was then
teaching at Einstein's old alma mater, the Eidgenössische Technische
Hochschule, where he remained until his death in 1958. Thus he was
forced to communicate with his colleagues by mail. Here is some of the
letter:

> Dear radioactive ladies and gentlemen,
>
> I have come upon a desperate way out regarding...the continu-
> ous spectrum, in order to save...the energy law. To wit, the possi-
> bility that there could exist in the nucleus electrically neutral parti-
> cles, which I shall call neutrons.

We have skipped the part of the letter in which Pauli described the
matter of the statistics. Note that he supposed that the particles—the
"neutrons"—carried no electric charge, since it was known that the observable daughter particles accounted for the total charge of the parent

nucleus. That notion has stood up; and despite all of highly charged—
in the emotional sense—theoretical speculation of the last few years, no
one has seriously proposed giving Pauli's "neutron" an electric charge.
The notion that the "neutron" must be *inside* the nucleus because it
comes *out* of the nucleus seems obvious enough. But, as we shall see,
when Enrico Fermi created it 4 years later, the modern theory of beta-
decay was built on quantum field theory in which particles need not
preexist but can be created out of the vacuum. No theorist would now
say that these Pauli "neutrons" preexist in the nucleus.

The next few sentences in Pauli's letter are most interesting because
the question raised in them has been at the forefront of a good deal of
the experimental and theoretical work in high energy physics and cosmology of the past few years, namely, does this "neutron" have a mass?
To the uninitiated, that might seem to be a foolish question, because, as
we look around us, *everything* from bees to bonnets seems to have a
mass. In fact, a Newtonian physicist would have claimed that everything
did have a mass. With Einstein's theory of relativity, however, the possibility that opened up was having energetic particles that were massless.
It followed from the theory that any such particle had to move with the
speed of light in a vacuum. Indeed, the above mentioned gamma-
particles—light quanta—are just such particles; that is, light quanta are
massless particles and, *a fortiori*, move with the speed of light.

What mass, then, should be assigned to Pauli's neutron? That is a
question to which we today would dearly love to know the answer. From
the energetics of the process it was apparent to Pauli, even in 1930, that
in the physicist's mass-energy units, the mass, or mass-energy, had to be
less than, or of the order of, the electron's mass. From now on, by
"mass" we shall mean "mass-energy," which, just to show how well some
quantities in physics can be measured, is given by 0.5110034 million
electron volt (MeV). That may be compared to the much larger mass of
the proton, which, in the same units, is 938.2796 MeV. (The absolute
significance of those units is not, for our purposes, as interesting as the
relative magnitudes. Just to keep things in perspective, however, 1 MeV
is equivalent to 3.8×10^{-14} calorie so that billions or even trillions of
electron volts are absurdly small amounts of energy when viewed on a
macroscopic scale.) This is what Pauli had to say about the "neutron's"
mass:

> The mass of the neutrons should be of the same order of magnitude
> as the electron mass and in any case not larger than 0.01 times the
> proton mass. The continuous β-spectrum would then become un-
> derstandable from the assumption that in β-decay a neutron is emit-
> ted along with the electron, in such a way that the sum of the ener-
> gies of the neutron and the electron is constant.

Some inkling of his state of mind when he made what was then the radical suggestion of accounting for an observable phenomenon by hypothesizing an unobservable one can be gleaned by the way Pauli ends the letter. He writes:

> For the time being I dare not publish anything about this idea and address myself confidentially first to you, dear radioactive ones, with the question how it would be with the experimental proof of such a neutron, if it were to have a penetrating power equal to or about ten times larger than a γ-ray.
>
> I admit that my way out may not seem very probable *a priori* since one would probably have seen neutrons a long time ago if they exist. But only he who dares wins, and the seriousness of the situation concerning the continuous β-spectrum is illuminated by my honored predecessor, Mr. Debye, who recently said to me in Brussels: "Oh, it is best not to think about this at all, as with new taxes." One must therefore discuss seriously every road to salvation. Thus, dear radioactive ones, examine and judge. Unfortunately I cannot appear personally in Tübingen since a ball which takes place in Zurich the night of the sixth to the seventh of December makes my presence here indispensable.... Your most humble servant, W. Pauli.

Pauli never did publish his speculation and, furthermore, never seems to have explored it quantitatively. It is one thing to make the qualitative suggestion that the emission of another particle in β-decay, along with the electron and daughter nucleus, would account for a continuous spectrum of electron energies; it is quite another to make a precise quantitative theory of the shape of that spectrum — to account for the relative number of electrons at each energy. This second step Pauli never took, probably because his own belief in his speculation was not strong enough to motivate him to do the necessary work. But he did talk about it. One of the people he talked to about it was Enrico Fermi, and it was Fermi who made the real theory. He also changed the name of Pauli's neutron. He decided that since *neutrone* in Italian means "big neutral one," the name "neutron" should be reserved for the particle, discovered by Chadwick, which has a mass of 939.5731 MeV, slightly heavier than the proton. He gave the name "neutrino" to Pauli's neutron, and that name stuck.

If Fermi had never done anything else in physics — something, given his genius and energy, that is unthinkable — the papers, and especially the 1934 paper in the *Zeitschrift für Physik*, on the theory of β-decay would have established him as one of the premier theoretical physicists of this century. In this paper Fermi applied, for the first time, notions of the then recently developed quantum field theory to what we would now call elementary-particle physics. In this theory, particles like the

·neutrino and the electron get created out of the vacuum and can disappear into the vacuum. In the β-decay, the decaying particle disappears into the vacuum and the emergent particles are created in its place. There is no question of neutrinos, for example, preexisting in the nucleus.

By using the formalism of the theory, Fermi derived a mathematical expression for the β-energy spectrum which, with subtle modifications, has endured to the present day. He allowed for the possibility — either possibility — that the neutrino could be massive or massless and showed that to be a testable proposition because the shape of the electron energy spectrum depends on the mass. That dependence, first written down by Fermi, has been used in the past few years to set an experimental limit — an *upper* limit — on the mass of the neutrino that is emitted in β-decay. The presently accepted limit is $m_{\nu_e} < 20$ eV, where we have put the subscript e on the neutrino to indicate that this is the neutrino that emerges in β-decay. We see that Pauli was mistaken when he thought that the mass of the neutrino might be of the same order of magnitude as that of the electron. It is, at most, 25 000 times smaller than the mass of the electron, and it might well be zero.

Pauli, in his letter, spoke of the difficulty of finding "experimental proof" for a particle with a "penetrating power equal to or about ten times larger than a γ-ray." It turned out that Pauli totally underestimated the dimensions of the difficulty. In his paper, Fermi does not discuss how one would go about detecting the neutrino. As was his wont, he focuses on the practical consequences of the theory. In particular, he points out how one could readily account for the fact that the observed lifetimes for the β-decays ranged from a few seconds to days. At least in part, that has to do with the energetics of the decay. He noted that all of this could be accounted for if a single "charge," or, as we now would say, "coupling constant" were assigned to the force responsible for the decays. It was not necessary to assign a different constant to each decay. Fermi's original theory contained the notion that all the annihilations and creations took place at a single point in space-time, something called a contact interaction.

As will be explained, that is not what we presently believe, and we regard Fermi's theory as a useful phenomenological representation of the deeper theory. For that reason, Fermi's charge, or coupling constant, had very peculiar dimensions, and it is not very enlightening to write the constant down. What was clear was that the force, or interaction, responsible for β-decay was something new. It was, on its face, much, much weaker than the electromagnetic force which determines the penetrating power of gamma-rays. To illustrate the difference, we all know that a window shade will block out light. However, we can readily calculate that if we took a *billion* earths and stacked them, end to end, a neutrino of the energy produced by the supernova would have only about one interaction in traversing this mass. It is little wonder that our earth is essentially transparent to neutrinos and that 30 million billion of them went through the much less massive IMB detector in 6 seconds and produced only 8 interactions.

For the next two decades following Fermi's paper, whatever progress there was in neutrino physics was in the theory. The notion of antiparticles was one of the things to emerge. Antiparticles have their genesis in two papers written in 1928 by the great British theoretical physicist P. A. M. Dirac, who, until his retirement, held the Lucasian chair at Cambridge University, which traces its origins back to Newton. Dirac's papers provided the first marriage between the theory of relativity and the quantum theory and showed that inevitable offsprings of such a union are antiparticles. At first those objects seemed more like love children than legitimate heirs, and for several years physicists scurried around trying to find some place to hide them, usually in the vacuum. In 1932, however, the Cal Tech cosmic-ray physicist Carl Anderson produced unimpeachable evidence, in cloud chamber tracks, that these particles actually exist.

What Anderson discovered was what we now call the positron, the antielectron. It has the same mass as the electron; and if it were unstable, which as far as we know it isn't, it would also have the same lifetime. But it does have an electric charge which is equal and opposite to the electron's. When an electron and positron meet, they can annihilate into γ-rays; "All the rest was γ-rays," as Harold Furth poetically described an imaginary encounter between Edward Teller and his antiparticle. Neutral particles may have antiparticles that are identical to themselves, the γ-ray being an example, or they may not. The neutron is an example of the second possibility.

The question of how to categorize the neutrino in this respect was first raised in 1937 by Ettore Majorana, an Italian physicist. Majorana, like the logician and computer scientist Alan Turing, was one of those extraordinarily gifted intellects whose psyches seem to be vessels too fragile to contain their intellects. Turing died in 1954 by eating a poisoned apple of his own devising, and Majorana disappeared at sea near Naples in 1938 at the age of 32 after having discussed the possibility of suicide. During his entire career he produced only nine papers, the one on the neutrino being the last. Fermi's group in Rome, of which Majorana was a member, bestowed various ecclesiastical names on one another. Fermi was known as the Pope for obvious reasons, and he, in turn, christened Majorana the Grand Inquisitor because of his deep and penetrating questions. Until Majorana's paper, it was taken as a given that the neutrino, like the neutron, was distinct from its antiparticle. That possibility—which may be correct, we still don't know—is known as the Dirac neutrino, and the opposite, the neutrino identical with the antineutrino, is known as the Majorana neutrino. It turns out that the modern theory of weak interactions makes the distinction irrelevant unless the neutrinos carry some mass. The questions of the neutrinos' masses and their Dirac or Majorana character are intimately linked.

The name "lepton" came into elementary-particle physics in the 1940s, and the original connotation from the Greek λεπτο'σ was "small" or "light." The first particles so designated were the electron, positron, and the neutrino and antineutrino. They were light in comparison with, say, the neutron or proton, which became known as baryons. If recent theoretical ideas are correct, there might be so-called leptons that are more massive than any baryon we have so far observed. Some of them might be created in the next generation of accelerators. They are nonetheless leptons in the now generally accepted meaning of the term; namely, they are particles which have only weak and electromagnetic interactions apart from gravity, which all particles have. Baryons, then, are particles which can have strong, nuclear interactions of which we will have more to say later, as well as the three other sorts already mentioned. Until recently, most elementary-particle physicists would have agreed that the lightest baryon—the proton—was absolutely stable, a mysterious fact which has, if it is true, no very satisfactory explanation. Indeed, as we have seen, the initial motivation for construction of the IMB detector was to test the proposition that the proton might *not* be absolutely stable. That proposition is usually called the conservation of baryon number; to each baryon is assigned a nonelectromagnetic charge called its baryon number. In conventional units, the baryon numbers of the particles we have considered so far are given in Table 1-1. Except where indicated, antiparticles have bars over their letters.

The assumption is then made that, to a very high degree of accuracy, perhaps even exactly, the baryon number is conserved in all reactions. In addition to their baryon numbers, particles are assigned so-called lepton numbers which also, empirically, are conserved to a very high

Table 1-1. Particles and Their Antiparticles and Their Baryon Numbers

Particle	Baryon number	Particle	Baryon number
p	1	e^-	0
$\bar{p}$	-1	e^+	0
n	1	ν_e	0
$\bar{n}$	-1	$\bar{\nu}_e$	0

degree of accuracy. In Table 1-2 we give the lepton numbers for the same group of particles as in Table 1-1.

Armed with these conservation laws, we can now reconsider some of the reactions already mentioned. The prototypical β-decay is that of the free neutron. It was not observed until 1948. The best present value of the neutron's average lifetime is 898 ± 16 seconds. Taking into account the conservation of lepton number, the decay is of the form

$$n \rightarrow p + e^- + \bar{\nu}_e$$

But all the variants, such as

$$e^- + p \rightarrow \nu_e + n$$

$$\bar{\nu}_e + p \rightarrow e^+ + n$$

$$n + \bar{p} \rightarrow e^- + \bar{\nu}$$

also satisfy the conservation of lepton number. The first two have been observed; the last has not been because we do not have an antiproton source of sufficient strength.

Testing of lepton conservation is going on actively at the present time. The trick is to find an unstable nucleus which cannot decay by the emission of a single beta-particle but which can decay if two particles are emitted. An example is ^{48}Ca, an isotope of calcium which can decay into ^{48}Ti, an isotope of titanium, with the emission of *two* electrons. If that were all that was emitted, one would have a violation of the conservation of lepton number which, according to present theoretical ideas, would then reveal both that the neutrino was Majorana and that it had a small mass. That mode of neutrinoless double β-decay has never been seen. There does, however, appear to be evidence for a decay mode in which two electrons and two antineutrinos are emitted. It does not violate the

Table 1-2. Particles and Their Antiparticles and Their Lepton Numbers

Particle	Lepton number	Particle	Lepton number
p	0	e^-	1
$\bar{p}$	0	e^+	−1
n	0	ν_e	1
$\bar{n}$	0	$\bar{\nu}_e$	−1

conservation of lepton number but is, nonetheless, very rare, because it involves two weak β-decay processes.

The Fermi theory was very successful in giving an account of the general phenomenology of β-decay—the raw physics—but there was something deeply unsatisfactory about it. Fermi had modeled his β-decay interaction as closely as possible on the quantum theory of electromagnetism: quantum electrodynamics. In that theory the photon—the electromagnetic quantum—acts as a carrier which transmits electromagnetic forces between electrically charged particles. The fact that this quantum has no mass, it was noted, accounts for the fact that the electrostatic force, that is, Coulomb's law, is what is known as long-range. Like gravitation, the electrostatic force between two charged point particles falls off as the inverse square of the distance that separates the particles. That is Coulomb's law, and it can be derived directly from the proposition that electrically charged particles exchange massless quanta. One may then ask, what corresponds to Coulomb's law for the weak force that is responsible for β-decay? In Fermi's theory the four particles involved in, say, neutron decay interact at a single point in space-time (Figure 1-1). This is to be contrasted to the electromagnetic force

Figure 1-1. The Fermi coupling.

which comes about with the exchange of a photon (Figure 1-2). In the original Fermi theory the weak force is of zero range, which corresponds to the exchange of a particle of *infinite* mass, a very unphysical proposition.

The first suggestion that the weak interaction might be transmitted by a particle of finite mass was made in 1934 by the Japanese physicist Hideki Yukawa. Yukawa's paper, which was published in the *Proceedings of the Physico-Mathematical Society of Japan*, is not, from our

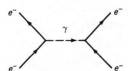

Figure 1-2. Electrons exchanging a photon.

present point of view, relevant in all its details, but it is one of those papers that sets the agenda for an entire field for decades. Yukawa was awarded the Nobel prize in physics in 1949 for his work. The main burden of his paper is not so much the theory of the weak interactions, which comes as sort of a bonus at the end, but rather to account for the force that holds nuclei together. At first sight it is remarkable that an atomic nucleus holds together at all. A nucleus contains protons, and they repel each other according to Coulomb's law. If that were the only force involved, the nucleus would fly apart. But it is known, mainly because of experiments done by Ernest Rutherford in the 1910's involving hard collisions of electrically charged particles with atoms, that the atoms have compact solid nuclei with dimensions of something like 10^{-13} cm, a unit of length we now call the fermi.

Yukawa was well aware of the connection between the mass of an exchanged particle and the range of the force that it generates. He noted that, if the range is taken to be 2×10^{-13} cm, a nuclear size, then the mass of the hypothetical carriers should be about 100 MeV. He called those carriers U-quanta, but the name was changed in subsequent years first to mesotrons and then, finally, to mesons. We have now introduced three categories of particles (baryons, leptons, and mesons) along with the electromagnetic quantum, the photon. There will be more.

In the last section of his paper, Yukawa turned to the subject of β-decay. To understand what he proposed, it is useful to compare the two parts of Figure 1-3. The first represents a meson exchange that contrib-

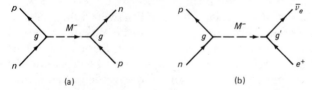

Figure 1-3. Meson exchange diagrams. (a) Nuclear force; (b) β-decay.

utes to the nuclear force and the second a meson exchange that contributes to β-decay. Note the two coupling constants g and g'; g is the strong coupling constant, and g' is the weak coupling constant. Although Yukawa did not discuss it, this theory makes a prediction for the reaction

$$\nu_e + e^- \rightarrow \nu_e + e^-$$

as Figure 1-4 makes clear. The prediction is that this reaction involves the coupling constants g'^2 as opposed to gg' for β-decay and g^2 for the strong force. In other words, neutrino electron scattering—collisions—should, according to the theory given by Figure 1-4, be several orders of

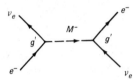

Figure 1-4. The scattering of electrons and neutrinos in Yukawa's theory.

magnitude less probable than β-decay phenomena. The prediction is false, but it is quite likely that if anyone had told Yukawa in 1934 that such collisions were going to be observable with powerful beams of energetic neutrinos of the sort first generated in Brookhaven in 1961, he would have thought that such an idea was quite mad. When Yukawa wrote his paper, no one had seen a single neutrino.

In the best of all possible worlds one might imagine that Yukawa's paper would have been widely read and would have inspired a search in cosmic rays for his meson. Unlike the neutrino which interacts only weakly, Yukawa's particle, the meson, would readily show up in cosmic-ray tracks. In fact, for several years Yukawa's paper was hardly known in the United States, and it had no influence on the discovery of what appeared to be his meson. (Carl Anderson, who discovered the positron and found the first evidence for what seemed to be Yukawa's meson, noted that, although he was generally familiar with it, Dirac's paper had no influence on his discovery of the positron, which was "accidental.") In the late 1930s a new penetrating component of cosmic radiation was discovered, and it was associated with a particle of a mass of about 100 MeV. In 1937 Robert Oppenheimer and Robert Serber made the suggestion that this object might, in fact, be Yukawa's particle. Over the next few years, however, a combination of theoretical and experimental work began to suggest that their idea was seriously wrong. Cosmic-ray experiments showed that the newly discovered particle was unstable. That was to be expected from Yukawa's theory (Figure 1-5).

The lifetime could be computed by using the value of g' gotten from β-decay experiments. When that was done, the calculated lifetime appeared to be some two orders of magnitude shorter than the observed lifetime, which was about 2 microseconds. To add to the puzzle, the cosmic-ray particles, which should, according to Yukawa's theory, have been strongly absorbed in matter, were behaving very strangely. The

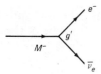

Figure 1-5. The decay of the M^-
in Yukawa's theory.

negatively charged ones were absorbed in lead but not in carbon. The experiments showing this were carried out in wartime Rome, under very difficult conditions, by the Italian team of Marcello Conversi, Ettore Pancini, and Oreste Piccioni and published in the United States in 1947. The experiments essentially ended the notion that this cosmic-ray particle was Yukawa's meson.

The way out of the dilemma was suggested in June 1947 at the first of the Shelter Island conferences which were international meetings of the then relatively few practitioners of elementary-particle physics. The Shelter Island conferences were succeeded by the Rochester conferences, which are now held all over the world and involve hundreds, if not thousands, of people. At the 1947 conference, Robert Marshak made the ingenious and, as it turned out, correct suggestion that what was being observed in the cosmic rays was the "daughter" of Yukawa's particle. He proposed that Yukawa's particle, which we now call the pi-meson, decayed into the particle the cosmic-ray experimenters were actually observing. This secondary particle acquired the name "mu-meson," although it is not a meson in the presently accepted sense of the term. In fact, in an act of prescience, Carl Anderson gave it its appropriate name when he first found it in cosmic rays. He called it a heavy electron, and that is exactly what it is.

The pi-meson, which was first found in cosmic radiation in 1947 by C. M. G. Lattes, G. P. S. Occhialini, and C. F. Powell, has all the attributes of Yukawa's particle. It is captured by the nuclei of the material through which it travels and in doing so releases a substantial amount of energy. The negative mu-meson is captured only in heavy materials like lead; in light materials it retains its identity long enough to allow it to decay. Hence the results observed by the Italian group. Into what does the pi-meson decay? Thereon hangs yet another tale.

By 1949 it was known that the decay of the muon, as the mu-meson is now known, was not a two-body decay. The visible decay product was an electron or a positron, depending on the charge of the parent muon. Just as in β-decay, this electron or positron demonstrates a spectrum of energies. Hence there must be at least two other particles emitted with it in the decay. Since they were not observed, it was natural, following

Pauli's original idea, to suppose that they were neutrinos or anti-neutrinos. In fact, to conserve lepton number, the decays must be of the form

$$\mu^{\pm} \rightarrow e^{\pm} + \nu + \bar{\nu}$$

It was realized over the next several years that this decay and the capture process given by

$$\mu^- + p \rightarrow n + \nu$$

were taking place at rates that were comparable to those of the corresponding electronic processes such as

$$e^- + p \rightarrow n + \nu_e$$

This meant that there was something universal about the weak interactions. The word "universal" is used in this context in the way in which it is used for gravitation or electromagnetism. We know that individual objects such as Mount Everest and a Ping-Pong ball, on a phenomenological level, exhibit very different gravitational manifestations. Mount Everest, for example, will deflect a surveyor's plumb bob. Nonetheless, we claim that gravitation is universal. What we mean is that any two comparable masses will be subjected to the same gravitational forces. Likewise, any two comparable electric charges will be subject to the same Coulomb force. In analogy, the idea began developing at about this time that there was one weak charge g and it was the same in all the weak processes, not just in different β-decays. The idea had deep reverberations in the next few decades.

In 1956, some 36 years after he invented it, Pauli's neutrino was finally observed. To put it more precisely, the antineutrino associated with β-decay, the object we have been calling $\bar{\nu}_e$, was observed. The achievement was the result of a beautiful experiment performed by the late Clyde L. Cowan and Frederick Reines, who were at the time members of the Los Alamos Scientific Laboratory. (Reines, incidentally, was one of the creators of the underground neutrino laboratory which now houses the IMB detector.) What made the original neutrino experiment possible was the existence of very large plutonium-producing reactors run by the then Atomic Energy Commission. The reactors, it turned out, are "factories" for producing antineutrinos. The reason why antineutrinos are produced is that, when nuclear fission takes place, the unstable isotopes that result are "neutron-rich": they have more neutrons than protons in their nuclei. They then begin to β-decay—to move in the direction of the stable isotopes, which have about the same number of neutrons as protons. The process is then a series of neutron β-decays which produce antineutrinos:

$$n \rightarrow p + e^- + \bar{\nu}_e$$

The successful experiment to detect antineutrinos was carried out at the Savannah River Plant in South Carolina in 1955–1956. The reactor produced a flux of some 5×10^{13} antineutrinos per square centimeter per second. By comparison, the supernova SN 1987 A generated, at the earth, about 10^9 neutrinos and antineutrinos per square centimeter per second. The supernova produced essentially an equal number of neutrinos and antineutrinos. It was fortunate that the reactor produced antineutrinos, because they can, in turn, generate positrons by the reaction

$$\bar{\nu}_e + p \rightarrow e^+ + n$$

This, incidentally, is the reaction that predominated in the supernova detectors, although the detectors also saw a small amount of the reaction

$$\nu_e + e^- \rightarrow \nu_e + e^-$$

In their experiment, Cowan and Reines made use of two metal tanks, the detector tanks, about 3 inches high and 6¼ by 4¼ feet wide. The tanks contained a solution, about 200 liters worth, of water and cadmium acetate. Since the experiment was done aboveground, it had to be shielded from cosmic rays. One could check on the validity of the shielding by seeing if there were events when the reactor was turned off. The real art of the experiment was in the selection of the signal which demonstrated an antineutrino absorption. It was there that the positrons played a crucial role. If a positron annihilates with an electron when it is at rest, two gamma-rays — very occasionally three — are produced. They emerge in opposite directions, and each carries an energy equal to the electron's rest mass, that is, 0.51 MeV. It is these gamma-rays one wants to detect. To see them, the detector is sandwiched between two layers of scintillating material which can record gamma-rays (Figure 1-6). The cadmium in the water is an absorber of neutrons; and when a neutron is absorbed, it gives off a burst of gamma-rays.

Now we can appreciate the ingenuity of the scheme; again see Figure 1-6. At a time which we can call $t = 0$, an antineutrino enters the tank and converts a proton into a neutron. That produces a positron. About 10^{-9} second later, the positron finds an electron in the target material,

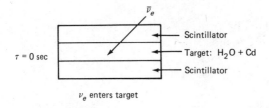

$\bar{\nu}_e$

$\tau = 0$ sec

— Scintillator
— Target: H_2O + Cd
— Scintillator

ν_e enters target

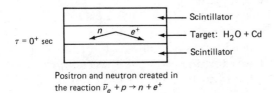

$\tau = 0^+$ sec

— Scintillator
— Target: H_2O + Cd
— Scintillator

Positron and neutron created in
the reaction $\bar{\nu}_e + p \rightarrow n + e^+$

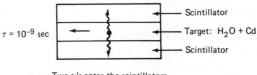

$\tau = 10^{-9}$ sec

— Scintillator
— Target: H_2O + Cd
— Scintillator

Two γ's enter the scintillators
after 10^{-9} seconds

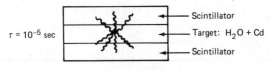

$\tau = 10^{-5}$ sec

— Scintillator
— Target: H_2O + Cd
— Scintillator

γ-burst entering the scintillators
after a neutron capture in cadmium

Figure 1-6. Sequence showing the time evolution of
the Cowan-Reines experiment.

with which it then annihilates. The resultant gamma-rays enter, in coincidence, the two scintillators. In the meanwhile the neutron created in
the neutrino absorption reaction has been wandering around the detector tank. In about 10^{-5} second it finds a cadmium nucleus into which it
is absorbed, making a gamma-flash. It is this sequence of events, spelled
out in Figure 1-6, that Cowan and Reines were looking for. The actual
experiments took several months to perform because there was an
antineutrino signal, even with the huge flux, of only about three per
hour. But the antineutrinos *were* seen, and so, after some 25 years,
Pauli's particle was discovered.

As we have seen, the neutrino was invented to resolve a paradox involving energy conservation in β-decay. We close this chapter with a second paradox involving neutrinos. This one is, perhaps, a little less evident, but as we shall see later, its solution will take us into one of the deepest puzzles in contemporary elementary-particle physics: Why are there families? Why, for example, is there an electron family duplicated mysteriously by a muon family, among others? We are still searching for an answer to the late I. I. Rabi's version of this question. When he first heard about the discovery of the muon, he asked, "Who ordered *that*?"

As we have seen, the muon decay is three-body, say,

$$\mu^- \rightarrow e^- + \nu + \bar{\nu}$$

On its face, however, there is no reason not to expect the decay

$$\mu^- \rightarrow e^- + \gamma$$

In fact, we can easily construct a theoretical diagram (Figure 1-7) which

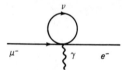

Figure 1-7. A possible μ-*e*-γ decay diagram.

produces it: This decay has never been seen. The present experimental limit is that there is at most 1.7×10^{-10} such decay per occurrence of the regular three-body decay mode indicated above. That inspires us to evoke what quantum mechanicians sometimes call the perfect totalitarian principle: That which is not absolutely forbidden is compulsory. In other words, if there is not some violation of a conservation law, then quantum mechanics insists that a process like the μ-*e*-γ transition must take place at some level. If it does not, there must be a reason. That is the paradox.

2

The Left
and the Right

In his rollicking autobiography, *Surely You're Joking, Mr. Feynman*, Richard Feynman describes how he got back into research in pure physics after having spent the war years at Los Alamos. He had just taken a teaching job at Cornell University. He writes,

> So I got this new attitude. Now that I *am* burned out and I'll never accomplish anything, I've got this nice position at the university teaching classes which I rather enjoy, and just like I read the *Arabian Nights* for pleasure, I'm going to *play* with physics, whenever I want to, without worrying about any importance whatsoever.

Feynman went to the cafeteria in Cornell and "some guy, fooling around, throws a plate in the air." He became intrigued by the motion of the plate, worked out the equations, and discovered an ingenious way of looking at the solution. He went to the man who had brought him to Cornell in the first place, Hans Bethe, with the solution. "He says," Feynman notes, "'Feynman, that's pretty interesting, but what's the importance of it? Why are you doing it?'" Feynman goes on,

> "Hah!," I say. "There's no importance whatsoever. I'm just doing it for the fun of it." His reaction didn't discourage me; I had made up my mind I was going to enjoy physics and do whatever I liked.

Fortunately for us, Feynman's interest moved on, soon after, from plates to quantum electrodynamics. He produced a remarkable series of papers which, at first, seemed very mysterious to most physicists because powerful results appeared to come out of nothing. By the late 1940s, Julian Schwinger had produced an apparently separate version

of quantum electrodynamics in which it was at least clear where the results came from, even if the mathematics was extraordinarily complicated. (At the time in Japan, unknown to everyone, Sin-itiro Tomonaga
had produced a third version of quantum electrodynamics which over-
laps with the others. The three men shared the Nobel prize in physics in
1965.) Soon after, Freeman Dyson showed how the different versions
were connected and made the field accessible to a large body of physicists.

Apart from the specific calculations in quantum electrodynamics that
he performed, Feynman invented a method, the Feynman diagrams,
which have since become the language of high energy theoretical physics. Several of the figures of Chapter 1, such as 1-1, 1-2, and 1-7, are
Feynman diagrams, although we did not so label them. What then is a
Feynman diagram? To begin with, no physically interesting quantum
field theory, such as quantum electrodynamics, produces field equations that can be solved exactly. The equations are solved in some sort
of sequence of successive approximations. As a rule, the approximations are given as a series—a sum of terms—which are labeled by suc-
cessively higher powers of a coupling constant such as the electric
charge e. If we call the function in the theory we want to know S, then
such a series can be written as

$$S = S^{(0)} + eS^{(1)} + e^2 S^{(2)} + e^3 S^{(3)} + \cdots$$

The Feynman diagrams, to which correspond precise mathematical expressions that can be computed by using the Feynman rules, represent
the various coefficients, $S^{(0)}$, $S^{(1)}$, $S^{(2)}$,..., and so on in the series. With a
little instruction, an otherwise innocent graduate student can, after
some practice, write the expressions down almost instantaneously. The
problem is that, in the theories that interest us, most of them are *infinite*. As they stand, they make no sense. That was the problem that confronted the theoretical physicists in the 1940s and still confronts us, although now we have a much better understanding of how to cope with
it.

The fact that quantum field theories were plagued with infinities was
well known in the 1930s. In our present terminology, there are two
classes of diagrams: loop diagrams and tree diagrams. In Figures 2-1
and 2-2 we give some examples of each class. The tree diagrams are always finite, but the loop diagrams are generally infinite. The diagrams
lead to what are known as radiative corrections, corrections to the tree
diagrams which, in the language of the expansion of the quantity S, correspond to the terms of lowest order. It can, and does, happen that the
tree diagrams give a very good account of many of the experimental

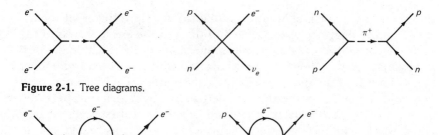

Figure 2-1. Tree diagrams.

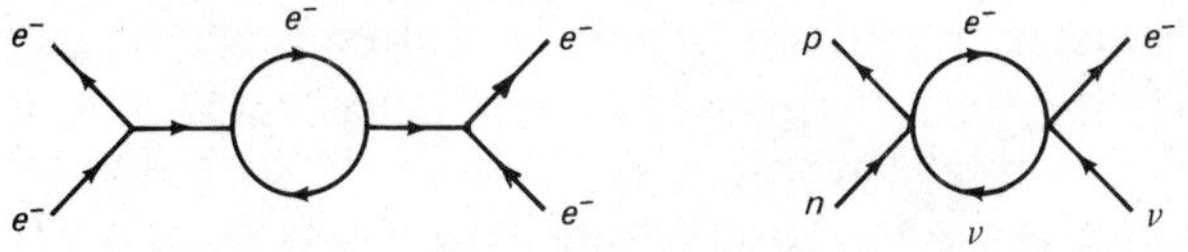

Figure 2-2. Loop diagrams.

facts not withstanding that, once one attempts to go beyond them, the theory fails to make sense because the terms become infinite. The Fermi theory of β-decay is precisely an example of that sort. Thus, in the absence of new experimental data, one was at a loss as to how to amend the theories.

In the late 1940s new data, at least in quantum electrodynamics, began to appear. The fact that it took until after World War II for these experiments to be done was not accidental. During the war, techniques for generating powerful beams of microwave radiation were developed. Microwaves have wavelengths of the order of centimeters and are used in radar, hence the wartime interest. Among the people working on radar was Willis Lamb. He got into the field a little late because, for a time, his foreign-born wife was considered an enemy alien. He was teaching at Columbia, and in 1944 he began working in the Columbia Radiation Laboratory, a sort of offshoot of the very large radar laboratory at M.I.T. Nonetheless, he continued to teach and, as it happened, he was assigned an atomic physics course for the summer session of 1945. While preparing his lectures, he began to realize that some of the wavelengths corresponding to interesting transitions among the energy levels of hydrogen were just in the microwave regime; they were the kinds of wavelengths that could be generated by devices, called magnetrons, which had undergone intense development during the war. At first, he thought all that he needed to do was to modify the existing magnetrons to produce the wavelength he wanted. In due course, he decided that he needed something else. Later he wrote, "Because I knew how to make only things that looked like magnetrons, my apparatus looked like a magnetron."

In 1946 Lamb managed to recruit a graduate student named Robert Retherford. Lamb later wrote, "He was an ideal partner for the project now increasingly dear to my heart." Lamb knew that, in the late 1930s, there were experimental indications that the standard theory of the hydrogen atom might be wrong. The positions of the energy levels of the hydrogen atom follow from quantum mechanics; they include the possibility that some of the levels might overlap with each other, something that is known as quantum mechanical degeneracy. The degeneracy in question, the one between the so-called $2p_{1/2}$ and $2s_{1/2}$ levels, persisted in the theory even if subtle relativistic corrections, first suggested by Dirac, were included. However, the prewar experiments hinted that, in reality, this energy overlap might be imperfect. A breakdown in the theory of the hydrogen atom is a very serious thing because, as the simplest composite system — one proton and one electron — it is the benchmark for the validity of the quantum theory. There were even suggestions that new, nonelectromagnetic forces might be at work.

The war then intervened and the matter was dropped. On April 26, 1947, Lamb and Retherford made their first measurement, and it became clear at once that the prewar results had been right except that now they were exhibited with very great accuracy. The shift of the $2s_{1/2}$ level above the $2p_{1/2}$ level is now known as the Lamb shift. It is usually given in frequency units, in this case, millions of cycles per second. A recent value, to give an idea of the accuracy, is 1057.862, with a small possible error in the last place. Lamb's first answer, presented in the Shelter Island conference of 1947, was about 1000 megacycles per second. This was an effect that could not be ignored.

It has been said that Hans Bethe did his profoundly important calculation of the Lamb shift on the train back from Shelter Island to Cornell. Given his facility, that would not be totally surprising. In any event, within five days of the end of the conference a preprint with Bethe's calculation was circulating. In it, Bethe made the first practical application of a technique which became, and still is, the key to useful, finite results from a formally infinite theory. It is the notion of renormalization. The quantum field theories, when written down, contain a variety of parameters, such as the masses and charges of the particles in the theory. The question is, should those parameters be compared directly with the measured values of those quantities? The answer is no. The theory provides a connection between the measured values and the parameters one begins with. For example, let us call the electric charge one begins with e_0. The theory then instructs us that the measured charge e is connected to e_0 by a relation of the form

$$e = Z(e_0)e_0$$

The problem is that when one tries to compute it by using Feynman graphs, Z turns out to be infinite in each order of e_0. The same is true of the mass and possibly other quantities. But to compare the results of any computation with experiment, we want to write the theoretical answers in terms of the physical observables such as e and m. That suggests shifting, or "renormalizing," the bare parameters of the theory to agree with the measured values. The process will get rid of at least some, and perhaps all, of the infinities. It was an idea which physicists like Kramers had already suggested in a general way. Bethe saw how to apply it to this case. He noticed that, if he shifted the electron mass, the worst of the infinities in the Lamb shift calculation could be absorbed. Bethe made some physically inspired guesses which enabled him to complete the calculation. He recognized that his calculation was not the final answer, since he had deliberately left out some of the relativistic effects. Nonetheless, he found an answer of 1040 megacycles per second, which was, in a certain sense, *more* accurate than Lamb's experimental result at that time; that is, it was closer to what is now believed to be the correct experimental answer.

At about the same time, other Columbia groups reported finding slight discrepancies in the magnetic properties of the electron which were accounted for in December 1947 by a virtuoso calculation done by Schwinger, inspired by Bethe's paper. As Schwinger later wrote,

> At the close of the Shelter Island conference, Oppenheimer and I took a seaplane from Port Jefferson to Bridgeport, Connecticut, where civilization, as it was then understood (the railroad) could be found. As the seawater closed over the airplane cabin, I counted my last remaining seconds. But, somehow, primitive technology triumphed. A few days later I abandoned my bachelor quarters and embarked on an accompanied, nostalgic trip around the country that would occupy the whole summer. [Schwinger went on his honeymoon.] Not until September did I set out on the trail of relativistic quantum electrodynamics. But I knew what to do.

Schwinger developed a fully relativistic formalism which was susceptible to renormalization and made the first truly relativistic calculation of radiative corrections. He presented a theoretical prediction for what is now known as the Schwinger magnetic moment of the electron. Ever since, there has been an interlocking series of theoretical and experimental results on this quantity as physicists have been testing, with greater and greater precision, renormalized quantum electrodynamics. Without worrying about the meaning of the quantity involved, but just to give a feeling for the detailed agreement, a recent experiment in which a single electron is trapped in an electromagnetic field for long periods of time yields

$$a = 1\ 159\ 652\ 200\ (40) \times 10^{-12}$$

where the parens indicate possible error. On the other hand, the theory gives

$$a = 1\ 159\ 652\ 460\ (44)\ (127) \times 10^{-12}$$

where the theoretical errors in the last places overlap with the experiment. Feynman diagrams which give these last places are so difficult to evaluate that much of the algebra, as well as the numerical analyses, is now done by computer. The computations have gotten beyond what a physicist can do unaided. Despite the infinities, quantum electrodynamics is, without doubt, the most accurate physical theory ever invented.

The success of quantum electrodynamics ushered in a renaissance in the study of the quantum theory of fields. Renormalizability became an essential criterion for a satisfactory field theory. What this means is that only a finite number of changes of scale are needed to render, term by term, the entire series of Feynman graphs finite. In terms of the scale factors, the Z's, renormalizability comes down to the proposition that only a finite number of the Z's are needed. In quantum electrodynamics, it turns out, only two different Z's are needed in addition to a renormalization of the electron's mass. The Fermi theory, in contrast, is "nonrenormalizable" in the sense that each order brings new kinds of infinities and no finite number of Z's will render the theory finite. Einstein's theory of gravitation is another example of a nonrenormalizable field theory.

In the early 1950s, however, not many people were interested in the problems the nonrenormalizable theories posed; that would change drastically a few years later. Instead, the attempt was to apply field theory methods to the strong nuclear interactions, to derive the nuclear force. Here the difficulties were of a different kind. In quantum electrodynamics, it turns out, the expansion of the various functions is controlled by the square of the electric charge. In suitable units this quantity, e^2, is about $1/137$; more exactly, $1/e^2 = 137.035\ 993\ (15)$ to quote a recent experimental value. With such a small parameter, one has every right to expect that successive terms in the series will become smaller and smaller. The question of whether the series as a whole converges to something is another matter. Even if it doesn't, as is likely in quantum electrodynamics, it may still be very useful. But in nuclear physics, the analogous coupling constants are, on their face, of order 1 or larger. It was not clear whether expanding in those constants made any sense, even if the underlying theories were renormalizable. Nonetheless, in desperation, people computed Feynman graph after Feynman graph, an exercise which in retrospect looks a little like the nineteenth century activity of measuring bumps on people's heads to determine either intelligence or the potential for criminality. At the time—that also would change a few years later—not a great deal of attention was paid to one of the most fundamental features of the theories, which was their ability to express symmetries.

What is a symmetry? Volumes have been written on the subject. For our purposes, a couple of concrete examples are probably more instructive than a good deal of abstract discussion. Imagine, as the first example, a sphere suspended in space. Our intuition tells us that there is something symmetrical about such an object. What we mean, in this case, is that if we rotate the sphere around any axis, it will not change its aspect. It will appear to be identical to the nonrotated sphere. This is an example of a continuous symmetry, since the rotations can vary continuously from a few degrees to a complete rotation or even several rotations.

In addition to continuous symmetries, there are discrete symmetries. As an example, suppose a collection of identical marbles in a box. To anyone presented with a second box identical except that the marbles are interchanged among themselves, the two boxes will look exactly the same. This is an example of the discrete symmetry known as permutation symmetry. Even prior to the invention of quantum mechanics it was understood that the great conservation laws of classical physics, such as the conservation of energy, momentum, and angular momentum, were connected to symmetries. For example, it was well understood that the conservation of energy was related to the fact that, in the description of events in which energy was conserved, the origin of time was arbitrary. What mattered was the *difference* between the times of the events. Likewise, the conservation of momentum could be related to the nonexistence of an absolute spatial origin.

With the invention of the quantum theory the connection between symmetries and conservation laws became even clearer. Moreover, the discrete symmetries came to play a very important role. One of the first significant applications of permutation symmetry in quantum theory was to electrons. When in classical physics we say, as in the example above, that a box contains two identical marbles, we really don't mean it. No two marbles produced by the hand of man are absolutely identical, but any two electrons *are* absolutely identical. They have the same mass, the same charge, the same spin—intrinsic angular momentum—and so on. Hence, if they are permuted among each other in a box, or an atom, the resulting physics should remain unchanged. That carried a remarkable implication, first postulated without proof by Pauli in 1925, before the invention of the quantum theory, that no two electrons could be placed in identical quantum mechanical states, something that is known as the Pauli exclusion principle. It underlies the periodic table of elements, since it determines how many electrons can be in any of the shells of electrons that surround the atomic nucleus. (Pauli was awarded the Nobel prize in physics in 1945, remarkably late in his career, for his exclusion principle, and not for his neutrino hypothesis.)

It was soon realized that the elementary particles could be divided into two classes: so-called Fermi-Dirac particles which obeyed the exclusion principle and so-called Bose-Einstein particles, which did not. (Satyendra Nath Bose was an Indian mathematician and physicist who discovered a new way of looking at the statistics of identical photons and sent his paper to Einstein, who immediately understood its deep meaning and developed the idea further. Bose-Einstein particles, of which the photon and pi-mesons are examples, can coalesce without limit in a given quantum state. In fact, that is what makes the laser possible; it is a coalescence of photons.)

Apart from permutation symmetry, other discrete symmetries began to enter quantum mechanics and then quantum field theory, whose development started in the late 1920s. These symmetries, it has turned out, have played an absolutely crucial role in the evolution of elementary-particle physics. Here we shall consider the three basic examples: charge conjugation, parity, and time reversal, beginning with charge conjugation. The notion of charge conjugation dates back to 1934, when Werner Heinsenberg stated his view that an essential feature of a quantum field theory was that the theory should be symmetric under the exchange of positive and negative charges. We now mean something more general, namely, a theory is said to be symmetric with respect to charge conjugation if it is invariant with respect to the exchange of particle and antiparticle. For many years it was thought that all theories had to have this property, that it was a law of nature and that, indeed, the universe as a whole had it. Quantum electrodynamics does have this feature, which accounts for important properties of the expansions of its functions.

The next discrete symmetry to be discussed is parity symmetry— sometimes referred to, rather imprecisely, as mirror symmetry. It is best discussed in terms of the two diagrams in Figure 2-3. A reader who is allergic to diagrams can reproduce them with the thumb and the first two fingers of the left and right hands respectively. Either of the coordinate systems, which are related by a reflection of the y-axis through the origin, can be used to locate objects in space. In fact, textbooks were written in either a right-handed or left-handed language and contained statements to the effect that the choice was conventional or arbitrary. Then came the Glorious Revolution of 1956.

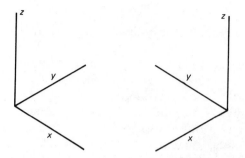

Figure 2-3. Right- and left-handed coordinate systems.

In our everyday experience most symmetries do not appear to be exact, and, without special pleading, most conservation laws seem to be approximate. If examined carefully enough, any sphere we can actually make, or draw, will have flaws that destroy its symmetry. A law of conservation, like that of the conservation of energy, seems to break down when we slide an object across the floor unless we take into account the vibrations of atoms in the floor that are excited by the dissipation of the energy of friction. Nonetheless, until 1956, nearly all physicists would have claimed that, just as the conservation of energy is a rigorous law of nature if we are careful in our accounting, so is the conservation of parity, which is another way of stating the symmetry between left and right. In fact, parity symmetry had been used successfully to make predictions in atomic and nuclear physics, and it was taken for granted that it would continue to be valid for the weak interactions such as β-decay.

In the early 1950s some paradoxical experimental results that began to appear at least suggested that the parity question for the weak interactions might be something worth examining. In the spring of 1956, Tsung-Dao Lee, of Columbia University, and Chen Ning Yang, then of the Institute for Advanced Study in Princeton, began such an examination. They did not really expect to find much room for parity nonconservation in the weak interactions because those interactions, especially the β-decay of nuclei, had been the subject of careful experiments for decades. Much to their astonishment, they discovered that none of the careful experiments that had been done had any bearing whatsoever on the question. That is a very important point. In fact, even if it had turned out that parity was conserved in the weak interactions, the work of Lee and Yang would still have been very valuable because it pointed out, for the first time, just which class of experiments does test which conservation law. In the three decades since Lee and Yang did their work, that way of thinking has become so ingrained in the ethos of elementary-particle physics that it is almost forgotten that someone had to think it up.

Results of the parity nonconservation experiments began coming in late 1956. In the meantime, Pauli had written to his former assistant, Victor Weisskopf, that "I am ready to bet a very large sum that the experiments will give symmetric results." They didn't. In fact, in a certain sense they gave the most asymmetric results possible. The first experiment completed was a joint Columbia–National Bureau of Standards project involving the β-decay of an isotope of cobalt, Co^{60}. The Co^{60} nucleus has an intrinsic angular momentum which can, with ingenuity, be aligned to a magnetic field; that is, the nucleus's own magnet can be aligned to an external magnet. If parity were conserved, it turns out, there could be no correlation between the direction in which the aligned angular momentum vector of the nucleus points and the direction in which the electrons produced in the β-decay emerge. In actual fact, as the experiment showed, there is a very strong correlation which establishes, definitively, that parity violation has occurred. Furthermore, experiments soon showed that charge conjugation invariance also was violated. Thus, two discrete conservation laws, whose validity had been thought to be absolute, were shown to break down. It was a stunning result.

Soon after these experiments became known, several theorists suggested a new view of the neutrino. It turned out that this "novel" neutrino theory had actually been in the literature since 1929, when it was invented by the mathematician Herman Weyl more as a mathematical curiosity than anything else. The theory appeared to violate parity conservation, so it had been abandoned. In describing it and his other work in physics, Weyl once commented, "My work always tried to unite the true with the beautiful, but when I had to choose one or the other I usually chose the beautiful." Without knowing of Weyl's work, but knowing that parity was violated by the neutrino-emitting weak decays, Weyl's theory was rediscovered.

The essence of the Weyl neutrino is that it has a "handedness." In less anthropomorphic terms, the neutrino, like the neutron and the electron, has an intrinsic angular momentum, a spin. In conventional units, the spin is ½. This means, for example, that a moving electron, depending on the circumstances, will have its spin pointing in some fraction of the cases in the direction of its momentum and in some fraction of the cases in a direction opposed to its momentum. On the other hand, the Weyl neutrino of β-decay always has its spin pointing in the direction opposed to that of its momentum. The antineutrino has just the opposite spin sense. See Fig. 2-4, in which the thin arrows represent the momentums and the thick arrows the spins.

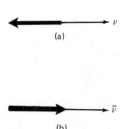

Figure 2-4. Spin and momentum of
(a) the Weyl neutrino ν of β-decay and
(b) the antineutrino $\bar{\nu}$.

Several remarks are in order. In the first place, since the neutrino is almost unobservable, how can we possibly determine empirically if such a property is actually realized in nature? Here the conservation of angular momentum plays a decisive role. The pi-meson, it turns out, has no intrinsic angular momentum. Hence when it decays, at rest, into a muon and a neutrino, the daughter particles must come out in opposite directions to conserve ordinary momentum. To conserve angular momentum if the neutrino is spinning to the left, the muon is spinning to the right. However, the muon interacts electromagnetically, and hence its spin orientation is something that can be and was tested. The result demonstrated to a high degree of accuracy that the neutrino emitted with the positive muon was left-handed and the object which we can call conventionally the antineutrino, and which is emitted in the decay of the negatively charged pion, is right-handed.

In a beautiful experiment done a little earlier at Brookhaven, Maurice Goldhaber, Lee Grodzins, and Andrew Sunyar had demonstrated that the neutrino emitted in β-decay is left-handed. It was very important to verify that both the neutrino emitted with the muon and the neutrino emitted with the electron are left-handed. It might, incidentally, appear that this fact has, by itself, settled the question whether the neutrino is a Majorana or a Dirac particle because, on its face, the particle we have labeled a neutrino has helicity opposite to that of the particle we have labeled an antineutrino. But for the massless Weyl neutrino theory that is really a convention. It ceases to be a convention if the neutrinos have mass. But then the matter can be settled only by the very difficult double β-decay experiments already alluded to.

Speaking of mass, it is absolutely essential to the Weyl theory that the neutrinos be massless. If they are not massless, and therefore not moving at the speed of light, it is possible to imagine an observer moving faster than they are. Such an observer would discover that the neutrino's spin orientation, with respect to its momentum, had turned around. The handedness of the neutrino would then be a property that varied from observer to observer. There is nothing wrong with that, except it is not the massless Weyl neutrino. That neutrino has a great deal of simplicity and elegance and may even be the one chosen by nature.

Pauli was stunned by these new developments. Recall that he had placed a sizable bet with Weisskopf that parity would turn out to be conserved. After the experiments indicated it was not, Pauli wrote a second letter to Weisskopf in which, characteristically, he raised the deepest question connected with the new discoveries. He wrote,

> Now after the first shock is over, I begin to collect myself. Yes, it was very dramatic. On Monday...at 8 P.M. I was to give a lecture on the neutrino. At 5 P.M. I received three experimental papers....I am shocked not so much by the fact that the Lord prefers the left hand as by the fact that He still appears to be left-right symmetric when He expresses Himself strongly. In short, the actual problem now seems to be the question: Why are strong interactions right-and-left symmetric?

In the thirty or so years since Pauli asked that question we still have not come up with a real answer. Nature seems to have apportioned out the discrete symmetries, parity and charge conjugation, in such a way that they are strictly preserved only by the nuclear and electromagnetic interactions and not by the weak ones. We can accommodate this in our theories, but accommodation is not the same thing as understanding. The situation with time reversal appears even more peculiar.

Classical physics made no particular explicit use of parity symmetry and no use at all of charge conjugation: Antiparticles are a quantum mechanical invention. It did, however, make use of time reversal invariance. That occurred in the great nineteenth century developments in kinetic theory and statistical mechanics, the study of the average behavior of ensembles of large numbers of particles. In considering elastic collisions among pairs of particles, it was explicitly assumed that the description of the collisions enjoyed what was known as microreversibility. The symmetry is best illustrated pictorially: Figure 2-5 illustrates the basic collision, but we can also imagine the *reverse* collision as illustrated by Figure 2-6.

The principle of microreversibility, which is what is meant by time reversal in classical mechanics, is the proposition that the reverse collision is just as legitimate a physical process as the original collision. If a film of these collisions were made, the reverse collisions would be made manifest if the film were run backwards. If we think about it, we soon realize that, in realistic situations, this notion appears to break down. For example, if we slide a realistic object across a realistic floor with fric-

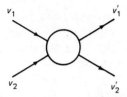

Figure 2-5. Two-particle elastic collision with the velocities indicated.

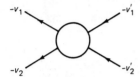

Figure 2-6. Two-particle collision the reverse of the one shown in Figure 2-5.

tion, the object will inexorably slow down. What then is the equivalent time-reversed motion and how would it be realized? It would be a motion in which the object would be reaccelerated. To analyze that, we would have to include an accounting of how, on an atomic level, the force of friction works, how the collisions between the atoms of the moving object and the atoms of the floor cause the latter to become stimulated, that is, heated up. The time reversal symmetry is realized only when we consider the full system of both the moving object and the floor and take into account the reversal of the atomic collisions.

In 1932, the Hungarian-born American physicist Eugene Wigner introduced time reversal into quantum theory. It was, in his words, "not quite trivial," which, translated, means that it was exceedingly subtle. Nonetheless, it became a standard tool in elementary-particle, nuclear, and atomic physics. By the time of the Glorious Revolution of 1956, a remarkable theorem had been proved by Pauli and others: the so-called CPT theorem. This theorem showed that, under very general circumstances, circumstances so general that they included any realistic quantum field theory, the combined transformation of charge conjugation C, parity P, and time reversal T had to be a symmetry of the theory. If T was a good symmetry, as everyone believed, then, even though P and C broke down in the weak interactions, the product CP also must be a good symmetry. That made the discoveries of 1956 seem a little less radical. The feeling lasted until 1964, when, in a stunning experiment done at the Brookhaven National Laboratories and to be sketched later, it was shown that CP invariance and thus, presumably, T invariance also, broke down.

One possible view is that, in addition to the four forces, strong, electromagnetic, weak, and gravitational, there is a "super weak" force which is not time-reversal or, equivalently, CP-invariant. Nonetheless, it is CPT-invariant, which guarantees that particle and antiparticle have the same masses and, if unstable, the same lifetimes. With this deepened understanding, an antiparticle must, strictly speaking, be defined as the CPT conjugate of a particle and not simply the C conjugate. If any experiment were ever to reveal, say, a small mass difference between an electron and a positron, indicating a breakdown in CPT, it would mean that an entirely new physical description of elementary particles would be required.

By the early 1960s, *beams* of neutrinos became available at Brookhaven. Considering that actual neutrinos had been observed by Cowan and Reines only a few years earlier, this development seemed almost incredible. The basic way in which such beams are produced now is the same as the one used originally at Brookhaven. With the old Brookhaven accelerator, protons were accelerated up to an energy of 33 billion electron volts, or 33 GeV. (G is for "giga.") These protons could be made to strike, for simplicity, a proton target: hydrogen. Inelastic proton-proton collisions of the form

$$p + p \rightarrow p + n + \pi^+$$

produce pi-mesons. Since they are charged particles, they can be guided, or beamed, by electromagnetic fields. In a neutrino beam-producing arrangement the pi-mesons are guided into what is known as a decay pipe, a long, partially evacuated tube within which the pions decay. The decay pipe presently in use at the National Accelerator Laboratory's Tevatron, the highest-energy proton accelerator in the world, which is located near Chicago, is 300 meters long, about the length of a football field. As one might guess from the name, the decay pipe is taken to be sufficiently long that, as they traverse it, essentially all of the pions decay. The principal decay mode of the pions, depending on their charge, is into either neutrinos or antineutrinos:

$$\pi^- \rightarrow \mu^- + \bar{\nu} \quad \text{or} \quad \pi^+ \rightarrow \mu^+ + \nu$$

The lifetime of the pions with either charge is about 2.6×10^{-8} second. The pions produced in the collisions are very energetic, which means that they are moving close to the speed of light. The speed of light is approximately 3×10^8 meters per second; hence, on its face, one would imagine that a typical pion would decay after traveling something like 8 meters. Why then is a decay pipe the length of a football field required? The answer has to do with the special theory of relativity. The lifetime quoted above would be the lifetime observed in a system *at rest* with respect to the pions. The theory of relativity teaches us that clocks appear to slow down when they are in motion. The lifetime is a form of clock, and the fact that it takes longer for the energetic pion sample to decay than would be expected is an excellent experimental confirmation of this aspect of the theory of relativity.

To see what use was made of the newly constructed Brookhaven neutrino beam in 1962, we return, once again, to the paradox we discussed at the end of the preceding chapter. It will be recalled that, to all intents and purposes, the decay $\mu^- \to e^- + \gamma$ appears to be absolutely forbidden. When something like that occurs in quantum mechanics, there is a reason. Following a nomenclature from atomic spectroscopy, such a reason is referred to as a selection rule, and it often reflects some deep symmetry property of the system. By the time the neutrino beam was being put in place, it had occurred to several physicists that the μ-e-γ paradox could be explained if there were the conservation of something that, for want of a better term, might be called "muness." To the negative muon we could assign a mu number of, say, 1 and to the electron a mu number of 0 and insist that the number be conserved in μ-decay.

That may seem entirely trivial until we realize what it implies. Let us look again at Figure 1.7, the Feynman diagram that leads to the unwanted μ-e-γ reaction. How is this diagram to be suppressed? It will be suppressed if we insist that it conserve muness at its vertices and if we assign to the neutrino in the loop the same muness quantum number as that of the muon. It then follows that this neutrino cannot be reabsorbed by the electron. But what this means is that we now have two "flavors" of neutrino: a muon neutrino and an electron neutrino. For example, we should be careful to write the decay of the muon as, say,

$$\mu^- \to e^- + \bar{\nu}_e + \nu_\mu$$

and the decay of the pion as, say,

$$\pi^- \to \mu^- + \bar{\nu}_\mu$$

But that apparently ad hoc conservation law—the conservation of muness—also makes a prediction, namely, that the reaction $\bar{\nu}_\mu + p \to e^+ + n$ is impossible, and the prediction can be tested.

Indeed, the neutrino beam at Brookhaven produced by the decaying pion, given this conservation law, was a beam of muon neutrinos or antineutrinos. The choice can be made by the experimenters by filtering out the pions of the "wrong" charge. That is what was done in 1962 by a Columbia-Brookhaven experimental group led by Leon Lederman, Mel Schwartz, and Jack Steinberger. (They were awarded the Nobel prize in physics in 1988 for their work.) Apart from the matter of selecting the right neutrino beam, there was the problem of shielding. The reaction allowed by the conservation law is

$$\bar{\nu}_\mu + p \rightarrow \mu^+ + n$$

Hence the target area must be shielded from muons that come from the original pion decays that produce the beam. In the event, the Brookhaven-Columbia group used steel plates cannibalized from an obsolete Navy cruiser to form part of the 44-foot-thick shielding. They were also able to make use of a then novel detection device, the spark chamber. In essence, the chamber consists of an array of parallel metal plates which have been electrically charged. When a charged particle passes between the plates, it causes the discharge of a series of sparks from plate to plate which, when photographed, shows the track of the particle passing through the chamber. Muon and electron tracks can be readily distinguished because of the great differences in the masses of the two particles.

The metal plates, aluminum in the original experiment, also serve as the target material for the neutrinos in the beam. In the original experiment some 10^{14} muon antineutrinos were produced in the beam per pulse of protons in the accelerator. The Brookhaven-Columbia group ran their experiment for 300 hours, but because of the very weak interaction of the neutrino, they were able to identify only 29 events. Every one of the 29 was consistent with the conservation of muness. There were now two distinct types of neutrinos, and, although no one realized it at the time, flavor had entered elementary-particle physics.

3

The Strange
and the Ordinary

At least during the past century, there has been an unusual affinity between physicists and mountain climbing. For most physicists, climbing is an avocation, something to take their minds off their troubles — troubles induced by thinking about physics. However, for the cosmic-ray physicists of the 1930s and 1940s it was a way of combining business and pleasure. A perfect example could be found in the person of the late Herman Hoerlin, who died in 1983 at the age of 80. At the time of his death, Hoerlin was in retirement from the Los Alamos Scientific Laboratory, of which he had been an important member since 1953. By his twenties, he had established a reputation as one of the best climbers in Europe and had made a number of first ascents in the Alps in winter.

In that era there were very few cable cars, so that even getting to the *base* of a high mountain in the winter was not trivial. Hoerlin was selected to be a member of a celebrated 1930, Swiss-led expedition to Kanchenjunga, an extraordinary 28 146-foot peak on the border between Nepal and Sikkim. The expedition failed, but as a consolation prize the group climbed Jonsong Peak, a 24 340-foot neighbor of Kanchenjunga which became for awhile the highest mountain in the world to have been climbed to its summit. In the meanwhile, Hoerlin had been studying physics at the institutes of technology in Berlin and Stuttgart. He received his doctorate at the latter institution in 1936. His thesis had to do with the effects of latitude on cosmic-ray intensity, a phenomenon connected to the earth's magnetic field. He made cosmic-ray measurements all the way from Spitzbergen, near the Arctic Circle, to the Straits of Magellan off southern Argentina.

In the course of his work, Hoerlin established what was then the highest cosmic-ray laboratory in the world at the summit of the Nevado Copa in the Cordillera Blanco in Peru at 20 300 feet. The reason for putting such a lab at high altitude was to minimize the background induced by the interactions with atmospheric particles. The first hint that there might be extraterrestrial cosmic rays came in 1910, when experiments were made at the top of the Eiffel Tower, about 984 feet above Paris. Existence of the rays was firmly established by Victor Hess the year afterward, when Hess made a remarkable manned balloon ascent to 17 552 feet. He is properly regarded as the discoverer of cosmic rays, and he shared the Nobel prize in physics in 1936 with Carl Anderson who, as we have mentioned, discovered the positron in cosmic rays.

By the 1940s, there were cosmic ray laboratories almost wherever there were mountains. Several could be found in the French Alps and American ranges like the Rockies, as well as the Andes. During World War II there were even B-29 flights at altitudes as high as 40 000 feet during which cosmic-ray experiments were done. As we have seen, the positron, the muon, and the pi-meson were all discovered in cosmic rays. However, nothing prepared physicists for what began occurring in the year 1946. Then two physicists from the University of Manchester, George Rochester and Clifford Butler, had a Wilson cloud chamber, a device for detecting ionizing radiation, set up on the summit of the Pic-du-Midi at nearly 10 000 feet in the Pyrenees. In October 1946, Rochester and Butler found a curious forked track in their cloud chamber. It looked something like the letter V. Since the track had no visible progenitor, they came to the conclusion that what must have happened is that a neutral particle had entered their chamber and then decayed into two charged particles. By working backwards, they concluded that what was, soon afterward, called the neutral V-particle had a mass somewhere between about 350 and 800 MeV.

A few months later, Rochester and Butler discovered another track that had a kink in it. This they interpreted to be a charged particle that had decayed into a neutral particle and another charged particle. Curiously, although these events were published, no particular notice was made of them. No doubt, that had partly to do with the fact that there was absolutely no "use" for any new particles. There was no obvious use for the muon either, but since the pion—the nuclear glue—chose to decay into it, we were stuck with it. There was simply no context into which to put a new particle. Furthermore, Rochester and Butler did not find any new V-particle events for nearly two years, giving the impression that the original events might have been a chimera. It later transpired that when cosmic-ray physicists who had been active with cloud chambers in the 1930s reexamined their old pictures they found them full of V-particle events which had been brushed aside. In science, as Pasteur was wont to observe, "chance favors only the prepared mind."

In the meantime, Cecil Powell and Giuseppe Occhialini, working at Bristol University in England, substantially improved a very important cosmic-ray detection device, the photographic emulsions. These were thick, dense, photosensitive materials within which a cosmic ray could stop and, in a manner of speaking, take its own picture. By 1950, a positively charged V-particle which decayed into three pi-mesons, π^+, π^+, π^-, with a mass between 435 and 495 MeV had been observed. It was given the name τ-particle. In addition, an entirely new object first discovered in a cloud chamber, which later became known as a Λ^0 particle, was observed to decay into a proton and a π^-.

If all of that seems to be confusing, it was. In particular, it was not clear whether each of the decay events of the original V-particle, or particles, was a new decay mode of a single object or the decay of an entirely new object. On the occasion of a noted conference in Bagnères-de-Bigorre in the Pyrenees in 1953, perhaps the high-water mark of cosmic-ray physics, the name K-meson was adopted for all particles with a mass intermediate between that of the pi-meson and the proton. Also, the name "hyperon" was adopted for any new particle with a mass between the mass of the proton and that of the deuteron, the nucleus of "heavy hydrogen," containing one proton and one neutron. Incidentally, the community of Bagnères renamed one of its squares the Place Hyperon in honor of the occasion. To use the modern nomenclature, the hyperons, nearly all of which were identified in Bagnères, and their approximate masses are given in Table 3.1.

Perhaps the most striking features of Table 3.1, apart from the bewildering diversity of a new particles revealed in cosmic rays, are the horizontal families. There are two families of two, n and p and Ξ^+ and Ξ^0, along with one family of three, Σ^-, Σ^0, and Σ^+, and the singlet Λ^0. Although physicists were not prepared for the influx of new particles, the arrangement into horizontal families did have a familiar ring. In 1938, one of Pauli's assistants, Nicholas Kemmer, introduced the mod-

Table 3-1. Hyperon Particle Table

Hyperons			Approximate masses, MeV
Ξ^-	Ξ^0		1320
Σ^-	Σ^0	Σ^+	1190
	Λ^0		1115
	n	p	940

ern version of what became known as isotopic spin. The empirical fact that inspired it was that the force, the nuclear force, that acts between a neutron and a proton is *grosso modo* of the same general magnitude as the force that acts between two neutrons or two protons. It is called the charge independence of nuclear forces. In addition to the nuclear force that attracts two protons to each other, there is also the much weaker Coulomb repulsion, the electrical force that would make a nucleus unstable if it were not otherwise held together by the much stronger nuclear forces.

Indeed, one could imagine a world in which the electromagnetic forces somehow got switched off. In that world, the notion is that the neutron and proton would become essentially indistinguishable. Two neutrons would, for example, experience the same attraction to each other as do a neutron and a proton. However, if we think about this in terms of Yukawa's meson theory—his original theory which involved only charged pi-mesons—we see that it is not possible to realize charge symmetry in it. What are missing are graphs of the form shown in Figure 3-1, where the particle being exchanged is a *neutral* pi-meson.

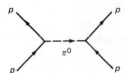

Figure 3-1. Two protons exchanging a neutral pion.

Kemmer predicted the existence of such a particle, and calculations done after World War II suggested that the particle would decay into two gamma-rays, that is, $\pi^0 \rightarrow \gamma + \gamma$, with a lifetime of about 10^{-16} second. In 1950 the π^0 was observed in experiments done with an accelerator at Berkeley, and hence it became the first of the new particles to have been discovered in an accelerator as opposed to a cosmic-ray experiment. When the newly discovered hadrons were found, and it was seen that they appeared to come in nearly mass-degenerate multiplets, it was natural to assign them isotopic spins also. In Kemmer's theory a triplet of degenerate mass states is given an isotopic spin 1, whereas a doublet like the proton and neutron has isotopic spin of ½ and a singlet like the Λ^0, as might be imagined, has isotopic spin 0. In this spirit it is natural to assign the Ξ particles isotopic spin ½ and the three Σ particles isotopic spin 1.

As we have mentioned, the 1953 Bagnères conference represented the high-water mark of elementary-particle cosmic-ray physics. To see why it did, we should realize that on a good day, on top of a fairly high mountain, a typical detector would turn up fewer than *two* V-particle decays. That was all right for qualitative studies, but if an experimenter wanted to know something to a few decimal places or wanted to know about something that was really subtle, as the *K*-mesons turned out to be, cosmic-ray techniques were all but hopeless.

Enter the era of accelerators. The notion of using electric and magnetic fields to accelerate charged particles so as to study their collisions goes back to close to the beginnings of nuclear physics. In 1932, John Cockcroft and Ernest Walton in Great Britain accelerated protons to several hundred thousand electron volts of kinetic energy and allowed them to bombard a lithium target. Thus, in a certain sense, they became the first people to "split the atom," that is, to induce a nuclear reaction with an accelerator. The Cockcroft-Walton accelerator was the prototype of what became known as a linear accelerator. Generically, such an accelerator consists of an ion source (a source of the charged particles to be accelerated), an evacuated tube in which the particles can travel while being accelerated by an electromagnetic field, and some means of producing the field. In the case of the Stanford linear accelerator (SLAC), the most powerful such machine presently operating in the United States, the tube is 2 miles long and klystrons accelerate electrons to fifty-odd *billion* electron volts. Klystrons are the microwave generators of the type that Lamb used in his Lamb shift experiment.

At the time of the original Cockcroft-Walton experiments, no one envisioned the technology—klystrons, for example—that would ultimately be required to go to really high energies in a linear accelerator. Indeed, it was thought that a few million electron volts was about the limit. In fact, if there is one thing that has been characteristic of the development of particle accelerators, it is that each step seemed like a dead end until someone had a brilliant, unexpected idea which enabled workers in the field to take the next step. Some years ago, the science fiction writer Arthur C. Clarke formulated what he called Clarke's first law: "If an elderly but distinguished scientist says that something is possible, he is almost certainly right; but if he says that it is impossible, he is very probably wrong." Clarke's first law has had many applications when it comes to predictions about the possibility and utility of new accelerators. In any event, when—erroneously, as it happened—it seemed impossible to go further with linear accelerators in the 1930s, Ernest Lawrence had a brilliant new idea. It was to make the accelerators *circular*.

It was well known that a charged particle introduced into a uniform magnetic field would tend to move in a circular orbit around the field lines. Lawrence made a brief calculation, a freshman physics calculation really, that showed that the time it took for a charged particle to make one circular loop around the field lines was independent of the radius of the circle. The greater its mass the longer the particle took to go around the loop, but the time, or the number of revolutions per second, was the same for any circular orbit of a particle with a given mass. Thus was the principle of the cyclotron born.

Figure 3-2a shows the outlines of a cyclotron when viewed from

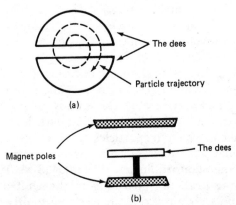

(a)

(b)

Figure 3-2. A cyclotron viewed (a) from above and (b) from the side with the dees between the poles of the magnet.

above. There are two pie-shaped evacuated chambers called dees, since, viewed appropriately, they resemble the letter D. In the middle, an electric field is placed across the dees. That is what accelerates the particles each time they come around a half circle. The polarity of the field must be changed so that the particles are always being accelerated as they cross the dees. In an ideal, classical cyclotron, and this is the significance of Lawrence's little calculation, the switch takes place, after the same time lapse, for each orbit no matter what its radius. That simplifies the engineering enormously. The dees, as Figure 3-2b shows, sit between the poles of the magnet that guides the particles in their circular orbits.

As a measure of the size of a cyclotron, the diameter of a magnetic pole face is usually quoted to give an idea of the size of the magnets used. The first model that actually worked, which was constructed in 1931 by Lawrence's student M. Stanley Livingston, had a 2.5-inch pole face diameter. It accelerated protons to 80 000 eV. It was soon followed by a 9-inch model that accelerated protons to a million electron volts (MeV), a real milestone. Lawrence was a very skilled promoter, as well as a brilliant scientist, and in the course of the next several years he raised the money to construct 11-, 27½-, 37-, and 60-inch machines, culminating in a 184-inch machine.

By 1937, however, Hans Bethe and Morris Rose had pointed out a limit to the classical cyclotron which, in turn, led to the next step in machine design. (The 184-inch machine, which was built in 1945, actually employed the new method.) Bethe and Rose noted that Lawrence's original cyclotron condition, the one that showed that the frequency of orbital rotation was independent of the orbital radius, had ignored the theory of relativity. That was all right as long as the proton kinetic energies remained at only a few million electron volts as compared to the proton's mass of nearly a billion electron volts. That is the mass of the proton when at rest, its rest mass. But the theory of relativity teaches us that particles become more difficult to accelerate when they are in motion, meaning that the effective mass of a moving particle, as measured by an observer at rest, increases with the particle's velocity. Bethe and Rose pointed out that this would destroy the simple relation between the frequency of orbital rotation and the radius of the orbit which had been the basis of Lawrence's original cyclotron design.

There were two ways to deal with the relativity condition, and each way led to a new generation of particle accelerators. In the Lawrence cyclotron the polarity of the electric field between the dees was switched rhythmically as the protons made a semicircular trajectory. This switching, in the original machines, could be made at a constant frequency, reflecting Lawrence's original observation that the orbital frequency, relativity aside, was independent of the orbital radius. One way to take into account the relativity effect was to change the frequency with which the polarity of the fields changed so as to keep in step with the proton orbits. An accelerator of that design was called a synchrocyclotron, and the first working example was the Berkeley 184-inch machine. By November 1946 it was producing 190-MeV deuterons.

Because the synchrocyclotron becomes too costly for energies in the billion electron volt regime, the second way of dealing with the relativity condition came into play, the so-called synchrotron. In the machines considered so far, the magnetic field that guides the particles in their circular orbits is fixed. The iron magnets that produce those fields got bigger and bigger as the energy of the machines increased, and before the invention of the synchrotron, they came to weigh in the hundreds of *tons*. In the synchrotron the magnetic fields are made to vary in time. The particles are accelerated in the fields until they are close to the speed of light. It then becomes increasingly difficult to accelerate them further. Since they are now moving with a fixed velocity, sensibly that of light, they will, in a given magnetic field, be contained in essentially a single orbit as opposed to the outwardly spiraling orbits of the Lawrence cyclotron, and they will remain in that orbit. In a synchrotron the particles move in a ring—literally an evacuated circular tube—and the guiding magnets, which can individually be much smaller than in the cyclotron, are placed around the ring.

In both of the new designs use is made of a crucial fact that was first noted by Vladimir Veksler in 1944 and then, independently, by Edwin McMillan in 1945. It is the so-called principle of "phase stability," which is the observation that particles in such a machine tend to travel in bunches. Suppose the protons, say, get out of step with each other as they move around the ring. Some will then arrive sooner and some later at the places where they are to be accelerated. Since the accelerating fields are changing with time, things can be accelerated so that a particle that arrives too soon will get a smaller boost than one that arrives on time. The effect on a particle that arrives too late will be just the opposite. The net result is that the particles get locked together in bunches. Unlike the classical Lawrence accelerator, which produced a continuous beam, the synchrocyclotron and the synchrotron produce particles in packets, packets that even in the early synchrotrons contained 10^{11} or 10^{12} protons each, and the packets arrive every few seconds. That is what put the cosmic-ray physicists out of the elementary-particle business except for the study of rare events at ultrahigh energies.

The first synchrotrons, which actually circulated electrons, began operating in the mid 1940s. By the early 1950s, the proton synchrotrons, which changed the face of elementary-particle physics, began coming on line. The first such machines were the 3 billion eV Cosmotron at Brookhaven, followed by a 1 billion eV machine in Birmingham, England, in 1953, and then the 6 billion eV Bevatron at Berkeley. The choice of 6 billion eV for this machine is significant. To produce an antiproton in a proton-proton collision, and to satisfy all the conservation laws, the reaction that requires the least amount of energy turns out to be

$$p + p \rightarrow p + p + p + \bar{p}$$

This minimum energy is essentially 6 GeV. Thus the bevatron was designed to discover the antiproton. This it did in 1955 in an experiment performed by Owen Chamberlain, Emilio Segre, Claude Wiegand, and Thomas Ypsilantis. On the other hand, pi-mesons had been observed in accelerator-induced collisions since 1948, and the proton synchrotrons produced K-mesons in such profusion that, unless they were shielded against, they swamped the detectors. Even prior to the advent of the proton synchrotrons, it was known from cosmic-ray studies that hyperons and K-mesons were being produced at rates that were at least a percent or so of the rates of pion production. On the other hand, these particles appeared to decay "slowly."

To understand what is meant in this context by "fast" and "slow," it is useful to describe a bit of the history of pi-meson physics after the accelerators began producing pions profusely. As we have mentioned in connection with neutrino beams, pions of the charged variety can be channeled into beams which can, in turn, be scattered off targets like protons. To give a measure of the strength of such a scattering, we define the notion of the cross section which the target particles present to the incoming beam. The cross section is a kind of effective area which the target particles block out. Physicists can study how this effective area changes as a function of the energy of the incoming beam; it is a little like studying how well a pane of glass filters light at different wavelengths.

In 1952 Enrico Fermi, using a synchrocyclotron at the University of Chicago, was studying the scattering of pi-mesons from protons in the region of about 140 MeV. He noticed that the cross section was rising rapidly. Some theorists had conjectured that something like this might happen but that the cross section should then turn over and trace a sort of bell-shaped curve. That would indicate a resonance, an energy regime in which pion-nucleon scattering was strongly enhanced. Fermi's machine did not have enough energy to observe the curve turnover, which later experiments showed would occur at about 180 MeV. At the time, no one thought of calling the resonance a new particle. It was usually referred to as an excited state of the proton in the way in which atomic physicists refer to energy levels of, say, the hydrogen atom as excited states. If we call this resonance, employing the modern term, Δ^{++}, then Fermi's reaction can be written

$$\pi^+ + p \to \Delta^{++} \to \pi^+ + p$$

Nowadays, it is customary to think of the Δ^{++} as an unstable particle. It is still Fermi's old resonance, but the attitude toward it has changed. Considered as a particle, the Δ^{++} does not have a definite mass. No unstable particle has a definite mass, which is a consequence of Heisenberg's uncertainty principle between energy and time. The more precisely the energy is known, the more time must the experiment that measures that energy consume. Here the shorter the lifetime the more uncertain the mass of the particle. In formula form,

$$\Delta m \, \Delta \tau \geq \frac{h}{2\pi}$$

where h is Planck's constant. The central mass of the Δ^{++} is now known to be about 1230 MeV. That is the mass at the center of the bell-shaped resonance curve, which itself has a width of about 115 MeV. By using the uncertainty principle, we can assign a lifetime to the Δ^{++} of the order of about 10^{-21} second. When we talk about a short lifetime, that is the sort of lifetime we shall mean. The interactions involved here are the strong pion-nucleon interactions, which is why the Δ^{++} decays very rapidly.

On this scale the hyperons and the K-mesons live for an eternity. The Λ^0, for example, lives about 2.6×10^{-10} second, and the charged K-mesons live about 1.2×10^{-8} second. Hence, two entirely different mechanisms seem to be at work in the production and decay of these particles. The production is rapid and the decay is slow. Moreover, certain selection rules became evident in the production. Attention was first called to them by Abraham Pais in 1951, following a suggestion by Oppenheimer, and they later became called associated production. To see what that means, consider two possible reactions:

$$\pi^- + p \rightarrow \pi^0 + \Lambda^0$$

and

$$\pi^- + p \rightarrow K^0 + \Lambda^0$$

The first is never seen; the second occurs copiously. That the new particles were produced together is what became called associated production.

The great step forward in thinking about the new particles came with the independent work of Murray Gell-Mann and the Japanese physicists Tadao Nakano and Kazuhiko Nishijima. It was the introduction of what Gell-Mann, with his knack for names, called strangeness. In what way were the new particles "strange," apart from the fact that no one expected them? Gell-Mann noticed that the isotopic spin assignments to the new particles were characteristically different from the previously familiar pattern. If we return to the conventional isotopic spin assignments and if we now give as a sublabel the value of the quantum number that specifies in which direction the isotopic spin is pointing, which we conventionally call T_3, we have

	π^+	π^0	π^-	p	n
T_3	1	0	-1	$\frac{1}{2}$	$-\frac{1}{2}$

This means we can write a simple equation that connects the electric charge of each particle to T_3 and the baryon number as defined in Table 1.1. To complete that table, we must now add the requirement that all mesons, pi and K, are assigned baryon number 0. The formula then says, where Q is the electric charge, that

$$Q = T_3 + \frac{B}{2}$$

We see, for example, that this correctly describes the fact that the charge of the neutron is 0 and that of the proton is 1. However, if we make the T_3 assignments for the Σ's and Λ^0

	Σ^+	Σ^0	Σ^-	Λ^0
T_3	1	0	-1	0

We see at once that giving all the hyperons baryon number 1, which we must do because we have baryon-number-conserving decays like $\Lambda^0 \rightarrow p + \pi^-$, makes the formula break down. It would, for example, give the Σ^+ particle the charge ½ rather than 1. Thus Gell-Mann introduced an additional quantum number, the strangeness S, defined in such a way that the formula

$$Q = T_3 + \frac{S}{2} + \frac{B}{2}$$

is generally satisfied. For the familiar particles—the neutrons, protons, and pi-mesons—the strangeness is 0, thus recovering the old connection between electric charge and isotopic spin. The Σ's and the Λ^0 are given strangeness -1, and the Ξ particles, which have isotopic spin ½, that is

	Ξ^0	Ξ^+
T_3	½	$-$½

are assigned a strangeness of -2. All the antiparticles have strangenesses which are opposite to those of the particles they are the antis of. For example, the $\overline{\Lambda}^0$, the antilambda, has strangeness $+1$. The K-mesons have the following isotopic spin assignments:

	$\overline{K}^0$	K^-		K^0	K^+
T_3	½	$-$½	T_3	$-$½	$+$½

which means that the K^0 and K^+ are assigned strangeness $+1$ and the $\overline{K}^0$ and the K^- are assigned strangeness -1.

What took this scheme beyond mere taxonomy was Gell-Mann's imposition of an approximate conservation law, namely, that in all strong processes, the kind in which the strange particles are produced, strangeness is conserved. That implies, in particular, the rules of associated production. For example,

$$\pi^- + p \to \Lambda^0 + \pi^0$$

violates the conservation of strangeness, whereas the reaction

$$\pi^- + p \to \Lambda^0 + K^0$$

conserves it. But strangeness conservation forbids the process

$$\pi^- + p \to \Xi^- + K^+$$

which associated production would allow. The fact that this reaction is not observed is a significant confirmation of the scheme. On the other hand, in the decays of strange particles, strangeness conservation breaks down as in

$$\Lambda^0 \to p + \pi^-$$

Hence these reactions, to use the old spectroscopic term, are "forbidden." They violate an approximate conservation law and thus take place at a slower rate. These rules do not "explain" the behavior of the strange particles, but they do serve to classify it.

Perhaps the most interesting developments in strange-particle physics that occurred after Gell-Mann's invention of strangeness were in the K-meson sector. First of all there was the so-called θ-τ puzzle, which first suggested that parity conservation in the weak interactions might be suspect. Until the experiments settled down, it appeared for awhile that there might be more than one charged K-meson, the θ and the τ. The so-called θ^+ particle had mass of about 494 MeV decayed into two pions:

$$\theta^+ \to \pi^+ + \pi^0$$

while the τ^+, which seemed curiously, to have a very similar mass, decayed into three pions:

$$\tau^+ \to \pi^+ + \pi^+ + \pi^-$$

It could, however, be shown that the two final states had opposite parities. But by 1954 it had become almost certain that the θ and τ had sensibly the same masses and lifetimes and that, in fact, they were the same particle: the object we now call the K^+. It is easy to see, in retrospect, that the way out of the dilemma of a single particle decaying into states of opposite parity was to have invoked parity nonconservation in the weak interactions that produce the decay. Until the work of Lee and Yang, however, that solution was assumed to be impossible because nearly everyone thought that the matter of parity conservation had been settled experimentally. Since we now know that parity is violated in the weak interactions, the θ-τ puzzle is no longer a puzzle; it is simply another example of parity violation.

In the spring of 1954, Gell-Mann gave a course of lectures at the University of Chicago on elementary-particle physics. Fermi was an auditor. After one of the lectures, Fermi asked a question the ramifications of which are with us still. He noted that both the K^0 and the $\overline{K}^0$ have a *common* decay mode; that is, K^0 or $\overline{K}^0$ goes into π^+ and π^-. But the K^0 and the $\overline{K}^0$ have opposite strangeness, and so the Feynman diagram (Figure 3-3) could interconvert them.

Figure 3-3. A Feynman diagram that mixes the K^0 and the $\overline{K}^0$.

Fermi wanted to know what consequences that might have. Upon returning to the Institute for Advanced Study, where he was then working, Gell-Mann, in collaboration with Pais, worked out the consequences. They were, and are, remarkable, an extraordinary example of quantum mechanical logic. To understand the logic, let us review what happens when a K^0 is produced in a reaction like

$$\pi^- + p \rightarrow \Lambda^0 + K^0$$

This is a strong interaction in which strangeness is conserved. Hence the object that is produced initially—the K^0—is in a state of definite strangeness, that is, strangeness $+1$. However, as this state evolves in time, it transforms into its antiparticle via Feynman diagrams like the one in Figure 3-3. This back-and-forth transforming continues until the particle decays. Thus the particle that decays is not in a state with a definite strangeness. In fact, since we begin with two states of definite strangeness, which can mix, we end up with two states *without* definite strangeness, each with its own lifetime for decay.

Furthermore, we are in luck, since the lifetimes of these states are very different. They differ by a factor of about 500. For that reason it is customary to call the state with the long lifetime K_L, for K long, and the other state K_S, for K short. It turns out that if all the processes that are responsible for this mixing preserve time reversal or, equivalently, CP, the states K_L and K_S have definite, and indeed opposite, symmetries under the operation of CP. That is essentially what accounts for the fact that the two states have such different lifetimes in the first place. The state we have been calling K_S is also the state that is symmetric under CP. Hence it can decay into two pi-mesons, which, in this case, also are symmetric under CP:

$$K_S \rightarrow \pi^+ + \pi^-$$

But the state that is identified as K_L is antisymmetric under CP and, if CP is conserved, it can decay only into final states that have a *minimum* of three particles such as $\pi^+ + \pi^- + \pi^0$ or $\pi^+ + \mu^- + \bar{\nu}_\mu$. These three-body states use up more of the available decay energy of the K meson than the two-body decays and, to use the technical phrase, leave less phase space for the decay. Other things being equal, decay modes of a particle into the fewest number of final particles will be the most rapid. That is why K_L and K_S have such different lifetimes. They also have different masses, but the mass difference is tiny, about 10^{-5} eV. As tiny as it is, it can be measured experimentally by studying in detail how the probability of finding a K^0 or a $\bar{K}^0$ in a beam that was initially composed of, say, a K^0 evolves in time.

All this was well understood in 1955, and for much of the next decade, it was taken as a given that CP was conserved and indeed that the behavior of the K^0–$\bar{K}^0$ system was a brilliant confirmation of CP conservation. Then came the shock. During the summer of 1963, an experimental group working at Brookhaven, led by Val Fitch and James Cronin of Princeton, observed two-pi-decay emanating from the long-lived K_L. This was a clear violation of time reversal, or CP invariance. The result was so shocking, and the experiment so difficult, that the experimenters spent nearly a year making sure that the effect was real before announcing it, a sort of object lesson in scientific restraint in our public-relations-oriented society. Cronin and Fitch were awarded the Nobel prize in physics in 1980 for this work. The original experiment has been repeated, by now, by many different groups with increasing accuracy.

Fortunately, these experiments are consistent with CPT conservation. Indeed, *all* experiments in elementary-particle physics are consistent with CPT conservation. The CPT symmetry remains valid even though C, P, and T break down individually. We still do not know why nature has ordered the discrete symmetries in this hierarchical manner, with more symmetries breaking down as the forces become weaker. When we turn to cosmology—the study of the universe at large—we will see that CP violation may be essential to understanding why the universe is, as far as we know, composed of matter with little or no antimatter. Perhaps mysteries that seem unresolvable in the microuniverse will yield when we consider the physics of the macrouniverse.

4
The Three and the Eight

In 1847, when Ernst Mach was 9 years old, he was asked to leave the Benedictine gymnasium near Vienna where he had been a student. The Benedictine fathers had decided that he was "*sehr talentlos* [hopeless]." As it happened, Mach's father, Johann, earned his living as an academic tutor. He proceeded to tutor young Mach at home, while shouting at him such imprecations as "Norse brains" and "head of a Greenlander." Even so, by age 15, Mach had made sufficient progress to return to the gymnasium and then go to the University of Vienna in 1855. In 1860, he took his doctorate in physics at the university, where he remained for the next few years while earning most of his living by giving lectures on physics to medical students.

Mach's research consisted, in part, of finding methods of demonstrating the Doppler shift in sound, which had been hypothesized by the Austrian physicist Christian Doppler in 1842. In an age when transportation faster than that of the horse was novel, the Doppler shift was still a controversial piece of physics. Indeed, one of Mach's own professors, Joseph Petzval, even claimed that the Doppler shift was impossible because it violated what Petzval had called the "law of the conservation of the period of oscillation." In 1878, eleven years after Mach had moved to the German university in Prague, he persuaded a group of students and professors to sit on a hill overlooking the train tracks and listen to the whistles of moving trains. Afterward they signed a document attesting to the fact that they had heard the Doppler shift.

As the lecture notes of his course for the medical students show — they were published in 1863 — Mach began his career as a dedicated atomist. He was thoroughly familiar with his contemporary Johann Loschmidt's statistical mechanical estimate of the size of air molecules (somewhat less than 10^{-7} centimeter in diameter) published in 1866. Indeed, Mach had presented papers to the Academy of Sciences in Vienna in which he used the notion of intermolecular vibrations to try to explain the spectra of gases. It seems to have been this work that led to Mach's disillusionment with the atomic hypothesis. Doing atomic spectra without quantum mechanics was like trying to construct a dictionary without a language. Mach was certainly not a physicist of sufficient distinction so that, by his scientific work alone, he would have had much influence on such of his great contemporaries as Max Planck, Ludwig Boltzmann, and, above all, Einstein. His influence developed because of his writing on the history and philosophy of science, and especially his great polemic work, *The Science of Mechanics*, published in 1883. Einstein, toward the end of his life, acknowledged that it had "exerted a deep and persisting influence on me." He spoke for a whole generation of physicists.

The burden of Mach's book was, in Einstein's phrase, the extermination of "harmful vermin." In this case the vermin were the metaphysical and theological assumptions that had crept into the foundations of Newtonian mechanics, with its unphysical notions of absolute space and time. The most famous passage of the book was Mach's critique of Newton's "proof," using the example of a rotating bucket full of water, that accelerated motions had absolute significance even in empty space. Mach noted that Newton's experiment proved nothing about rotations in empty space, since it was not done in empty space. Indeed, he pointed out, the same effect might well be produced by rotating the stars around the stationary bucket, something that Einstein kept in mind when he was creating his general theory of relativity and gravitation. When Mach turned against the atomic hypothesis, his prestige as a philosopher of science was such that he converted, at least for awhile, a good many physicists, including, incidentally, Planck.

As fate would have it, the year before Mach returned to the University of Vienna in 1895, in a chair of philosophy, Boltzmann had been appointed a professor of physics. Boltzmann, who was the foremost statistical mechanician in Europe, was a firm believer in the utility of the atomic hypothesis, so it was only a matter of time before he and Mach were at odds. Indeed, Boltzmann wrote,

> I once engaged in a lively debate on the value of atomic theories with a group of academicians, including Hofrat Professor Mach, right on the floor of the academy of science itself....Suddenly Mach spoke out from the group and laconically said: "I don't believe that atoms exist." This sentence went round and round in my head.

In debates like this, Mach was fond of asking *"Haben Sie einen gesehen?* [Have you seen one?]" This apparently reasonable question has once again come back to haunt us in the matter of the quark.

Viewed in retrospect, the steps that led to the quark appear to have all the inevitability of a solved crossword puzzle. But to anyone who witnessed the process, nothing about it seemed obvious. Indeed, much of it, to use Pauli's phrase, seemed like *"desparate Physik* [desperation physics]." Much of the desperation was caused by the apparently incessant discovery of new particles, most of which seemed to have no rhyme or reason. An exception was finding the first meson-meson resonances. In the spirit of the Δ^{++}, they were soon identified as particles, in this case the ρ and the ω. They had, to a degree, been anticipated. Since the mid-1950s, Robert Hofstadter, with a newly constructed 1-Gev linear electron accelerator at Stanford University, had been studying the collisions between electrons and various nuclei, including the proton. This turned out to be an immensely instructive enterprise, for which Hofstadter won the Nobel prize in physics in 1961. It had been anticipated that the strong interactions between neutrons and protons, and the various pi-mesons, would affect the electromagnetic properties of the particles concerned in a very substantial way. That is entirely analogous to the way radiative corrections affect the electromagnetic properties of the electron; the Schwinger magnet moment, for example, is such an effect.

The proton magnetic moment had been measured as early as 1933, and it differed very substantially from the value to be expected from the simple Dirac theory. In the language of the Yukawa theory of pi-mesons, this difference could, in a qualitative way, be attributed to Feynman diagrams such as Figure 4-1. A diagram like this suggests that

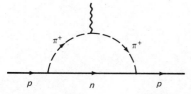

Figure 4-1. A Feynman diagram that contributes to the electromagnetic structure of the proton.

the proton has a "size." The argument uses the uncertainty principle between energy and time:

$$\Delta E \; \Delta \tau \; \simeq \; \frac{h}{2\pi}$$

If the pion shown being emitted in Figure 4-1 were "real"—had its physical mass—there would be a violation of the conservation of energy of the order of the pion mass, about 140 MeV. But that is allowed, according to the uncertainty principle, so long as the pion is reabsorbed rapidly enough. The uncertainty principle allows a time of the order of

$$\Delta\tau \simeq \frac{h/2\pi}{m_\pi c^2}$$

where c is the speed of light and we have used Einstein's formula $E = mc^2$ to fix the energy that goes into the uncertainty relation. But, if the pion moves with the speed of light (about 3×10^{10} cm/sec), it will have traveled a distance of about

$$c \, \Delta\tau \simeq \frac{h/2\pi}{m_\pi c}$$

from the proton before being reabsorbed. On putting in the numbers, we find about 1.4×10^{-13} centimeter for the quantity $h/2\pi/m_\pi c$, which is known as the pion's Compton wavelength. According to this argument, the Compton wavelength should be about the size of the proton's electromagnetic structure. Experiments done by Hofstadter suggested that the proton did, indeed, have a size of roughly that order of magnitude. But Hofstadter was also able to measure the electromagnetic size of the neutron. Even though the neutron is electrically neutral, it can still have an electromagnetic structure consistent with its charge neutrality. Figure 4-2 indicates how that can come about. When the ex-

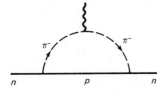

Figure 4-2. A Feynman diagram that contributes to the electromagnetic structure of the neutron.

periment was actually done, however, the neutron seemed to exhibit hardly any electrical size at all. That was extremely puzzling. A few theoretical physicists had the nerve to suggest that it might signal the existence of a class of new particles—pion-pion resonances that had spin 1 and were electrically neutral like the photon—whose effect on the electrical structure would be just such as to add positively for the proton and cancel out for the neutron. That seemed a little far-fetched, but it turned out that these theorists were exactly right.

In 1961 the first of these resonances was discovered at Brookhaven; it was the so-called neutral ρ-meson. It was a broad two-pion resonance with a central mass of about 770 MeV and a width of about 153 MeV. It was a chargeless, spin-1 object just as the theorists had guessed. Crucial to this discovery was the utilization of the bubble chamber. That wonderful device was invented in 1952 by Donald Glaser, who was then a postdoctoral fellow at the University of Michigan. (It was said that Glaser was inspired by watching bubbles form in bottles of beer.) The idea was to take a substance, in the first instance ether, and to suddenly reduce the pressure on it below that at which the substance normally boiled, something that is called superheating. The substance is thus put into an unstable state in which boiling will occur at the sites where an electrically charged particle has passed. The particle leaves a visible track in the chamber.

The first device in which Glaser actually observed tracks consisted of a small glass bulb 1 centimeter in diameter and 2 centimeters long filled with ether. Keep those miniscule dimensions in mind when we consider the more modern versions. No doubt, sooner or later, the utility of such a device for detecting elementary particles would have been widely appreciated and the device developed, but the fact that it happened sooner rather than later, and the colossal way in which it did happen, certainly had to do with the various skills of the late Luis Alvarez, who died in 1988. (Incidentally, Glaser won the Nobel prize in physics in 1960 for the invention of the bubble chamber and Alvarez, in 1968, for his contributions.) Alvarez immediately understood the potential of the device especially if it could be greatly enlarged and, more especially, if it could be filled with either hydrogen or deuterium, which would then provide relatively simple targets to analyze. He at once assembled a team to begin building bubble chambers, a team which included Glaser. They started by building chambers that were a few inches in diameter. Alvarez was sure that, to study the decay of strange particles, he would need a chamber that was at least 30 inches long. In the event, he decided to build a 72-inch-long chamber even though, at the time, the largest working chamber the group had was only 4 inches long. In his autobiography, *Alvarez*, he describes what sort of challenge the 72-inch chamber presented.

> [It was] to be 72 inches long, 20 inches wide, and 15 inches deep. It had to be pervaded by a magnetic field of 15 000 gauss, so its magnet would weigh at least a hundred tons and would require two or three megawatts of power to energize it. [The magnet is needed to bend the tracks of the charged particles whose orbits are influenced by the magnetic field. By measuring the curvature of those tracks,

information about the charge and mass of these particles can be gotten.] It would require a glass window 75 inches by 23 inches by 5 inches to withstand the operating pressure of eight atmospheres, a force on the glass of a hundred tons. [The glass window, which was exceedingly difficult to make, was necessary for viewing and photographing the tracks in the chamber.] Only the first hydrogen bomb builders had any experience with such large volumes of liquid hydrogen; the hydrogen-oxygen engines of the Apollo lunar rockets were still gleams in the eyes of their designers.

The 72-inch chamber, which cost $2.5 million in 1955 dollars, began operating in 1959, and, at about the same time, a hydrogen bubble chamber — the one in which the ρ^0-meson was discovered — began operating at Brookhaven. It was understood from the beginning that human, unaided scanners would not be able to process the hundreds of thousands of photographs of the events occurring in a large bubble chamber. At both Brookhaven and Berkeley, major advances in the automation of the scanning procedure made the discovery of something like the ρ^0 possible. Not long after ρ^0 was found, two new pi-meson resonances turned up at Berkeley: the ω^0 with a central mass of about 783 MeV and a very narrow width of only 9.8 MeV (with such a narrow width, it could really be interpreted as a particle that decayed into three pi-mesons) and the η^0 with a mass of about 549 MeV and a width of 1 KeV. The η^0, which, like the ω^0, decays into three pi-mesons, appeared at first to be another love child like the muon. It had spin 0, and that meant it had no role to play in the electromagnetic structure of the nucleons, since the photon has spin 1.

That was only the beginning. New resonances and/or particles began arriving from all directions. Some had strangeness. Some hadn't. Some were charged. Some weren't. Some had vanishing spin. Some hadn't. Some were hyperons. Some weren't. The whole thing seemed to be getting out of hand. Nature, in this corner, appeared to have gone slightly berserk.

In retrospect, one can begin to see three very significant theoretical themes emerging from the general fog. At the time they were introduced, it was not clear to anyone, including, one would imagine, their proponents, what they would lead to. The first idea was due to Fermi, who, with Yang, worked it out in detail in 1949, before the real advent of the strange particles. Fermi noticed that if one bound a proton and an antiproton together in a state of zero angular momentum, the product would, as far as its general properties (spin, charge, and parity) were concerned, be indistinguishable from a neutral pi-meson. Hence, he and Yang introduced the idea that pions — it also works for the charged pions — were, in "reality," bound states of nucleons and antinucleons.

The virtue of the Fermi-Yang scheme was that it reduced the number of fundamental or "elementary" particles. That had the sort of reductionist appeal of nuclear physics in which one says, for example, that a helium nucleus is, in "reality," two neutrons and two protons bound together or of atomic physics in which one says that the hydrogen atom is, in "reality," an electron and a proton bound together. There was, however, a monumental difference: the energies involved. When two particles are bound together, a certain amount of energy is lost. We know that because, to pry them apart, we need to supply energy. In each of the examples given, the amount of energy lost, the so-called binding energy, is a small fraction of the rest masses of the particles being bound together. For example, the binding energy of the electron and proton in a hydrogen atom is 13.6 eV and the combined rest masses are about 939 MeV, which means that the mass loss due to binding is a little more than 10^{-8}th of the rest masses involved.

On the other hand, if we put a proton and an antiproton together to make a pi-meson, since the proton and antiproton have individual masses of about 938 MeV and the neutral pion about 135 MeV, the percentage mass loss is nearly *93 percent*! This means that the binding must be extremely strong, and no one had any idea of how to deal with such a regime. In the interest of reducing the number of elementary particles, an intractable dynamical problem apparently had been introduced. Nonetheless, when the strange particles began to appear, this scheme was elaborated by adding strange particles to the mix. The most successful of these models was due to the Japanese theorist Shoichi Sakata. Published in 1959, it combined the proton, neutron, and the Λ^0 with their antiparticles to reproduce the other then-known mesons and hyperons. Although the details of the model are not, from our point of view, correct, it was the first time that a model in which triplets of fundamental particles played a key role was tried, a foretaste of things to come.

The second theoretical theme to emerge at this time was the pursuit of what became known as higher symmetries. "Higher" as compared to what? We have already, in Chapter 3, discussed the role of isotopic spin in the classification of particles. Even among the newly discovered resonances, the pattern of the isotopic spin multiplets continued. For example, the ρ's came in three varieties, ρ^+, ρ^- and ρ^0, with comparable masses, whereas K^*'s—a K-π resonance with a mass of about 892 MeV—came in isotopic doublets like the K-mesons. If these isotopic multiplets can be thought of as a "lower symmetry," then by a "higher symmetry" was meant a symmetry more embracing than isotopic spin, something that would unify strange and nonstrange particles. Why was that so difficult to find? Let us look a little more closely, as an example, at the pi-meson isotopic spin triplet, this time indicating the true measured masses.

We see from Figure 4-3, recalling that the square of the electron charge e^2, in suitable units, is about $1/137$, that the mass difference among the pions—charged and neutral—is of the order of $e^2 m_\pi$. (The π^+ and π^-, being antiparticles of each other, must, by the CPT theorem, have exactly the same mass.) We do not believe that is a mere numerical coincidence. We think it reflects the fact that it is the electromagnetic interaction that breaks the mass degeneracy among the pions. If we look at the other mass splittings among the isotopic multiplets, all of them are of the order of $e^2 m$, where m is the central mass. That further encourages us in our view of the electromagnetic origins of mass splittings.

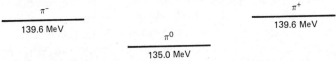

Figure 4-3. The pion isospin triplet showing the mass splitting.

But, to take an example, the K and π are split in mass by 355 MeV, over twice the mass of the pion. Given this mass splitting and the other apparent dissimilarities between K and π, it was difficult to imagine *a priori* how the two could be united in any kind of multiplet structure. In the late 1950s, heroic efforts were made to try to find some unifying principle, but they did not seem to be leading anywhere.

The third theoretical strand, first introduced in the mid-1950s, goes under the daunting rubric "non-Abelian gauge theories." It has turned out to be the key to everything, although almost no one took it seriously at the time. In classical electrodynamics, the quantities that determine the behavior of the charged particles are the electric and magnetic fields. To find them, however, an auxiliary quantity called the electromagnetic potential must first be introduced. This quantity is a function of space and time and is a vector in the four-dimensional space of Einstein's theory of relativity. If it is known, it is, at least in principle, straightforward to compute the electric and magnetic fields. But it turns out that there is an arbitrariness in the definition of the potential. Indeed, there are an infinite number of potentials, connected to each other by the so-called gauge transformations, that produce the same electric and magnetic fields. Thus, to begin solving the Maxwell equations, one must choose one of the potentials out of this class: select a gauge. This freedom is welcome, since different gauges may be more or less suitable for different problems.

When quantum electrodynamics was created in the late 1920s, it was understood that gauge invariance would have to be built into the theory. In the theory the electromagnetic potential of Maxwell becomes identified with the quantum field of the photon, the mathematical object that creates and annihilates photons. If this sector of the theory is modeled on the classical version, it is, just as the classical theory was, manifestly gauge-invariant. However, quantum electrodynamics is physically interesting only when photons are allowed to couple to charged matter like electrons and positrons. Then the concept of the gauge transformation must be enlarged to include those matter fields.

How to do that was well understood in the 1930s, but over the years, a different way of looking at it developed. It was realized that the principle of gauge invariance actually determined how electrons and photons interact. If it were invoked, one could rule out non-gauge-invariant photon-electron interactions, which, in any case, were not observed. Moreover, the principle of gauge invariance gave an explanation of why the photon has no mass, that is, why the Coulomb force was long-ranged. It was easy to see that putting a photon mass into the equations, which would make the Coulomb force short-ranged, would also spoil the gauge invariance.

In the early 1950s a few physicists began to explore generalizations of the gauge symmetry which might have some bearing on the strong interactions. Pauli made an attempt but hit a snag, of which more shortly, and never published what he had done. In 1954 a doctoral student at Cambridge University named Ronald Shaw wrote a dissertation on these gauge theories that was never published, and a prescient paper had been published in 1938 by the Swedish physicist Oskar Klein but had been totally forgotten. The paper of monumental importance on this subject, that of Chen Ning Yang and Robert Mills, was published in 1954. It is hard to think of a postwar paper in elementary-particle theoretical physics that has had more significance although, at the time, it was almost completely ignored. Yang and Mills imagined a world of strongly interacting particles—only neutrons and protons in their paper—in which electromagnetism was somehow shut off. In this world the neutron and proton are physically indistinguishable; it becomes a matter of convention which particle we call which. Prior to the Yang-Mills paper the assumption was that, once this choice had been made at a given point in space and time, the *same* choice should hold everywhere in space and time. That is something we can call global isotopic spin invariance. Yang and Mills proposed to generalize this idea so that a *different* choice of neutron and proton could be made at each point in space and time, something we can call local isotopic spin invariance. If we demand that a field theory satisfy this condition, there are very clear consequences.

Just as the demand that quantum electrodynamics should be gauge-invariant brought forth the photon, so the demand that the nucleons participate in a locally isotopic-spin-invariant theory brings forth an analog of the photon. It turns out that what is actually brought forth here are three "photons"—spin-1 quanta—which Yang and Mills called *b*-quanta. These *b*-quanta differed from the old-fashioned photon in one very important respect: They interacted directly with each other. That turned out, and again no one realized it at the time, as we will discuss later, to be such an important point that it is worth illustrating it with a Feynman graph showing four *b*-lines meeting at a point, a diagram that would contribute to the scattering of *b*'s from each other (Figure 4-4).

Figure 4-4. The scattering of *b*-quanta at a point.

In contrast, the electromagnetic photons can interact with each other only through the mediation of charged particles. Figure 4-5 is the famous box diagram for photon-photon scattering by the exchange of electrons. These direct self-couplings of the "photons," now generally

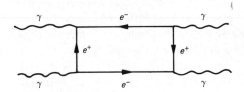

Figure 4-5. Photon-photon scattering via the exchange of electrons.

called gauge fields, are characteristic of non-Abelian gauge theories of which the isotopic spin gauge theory is the simplest example. (The term "Abelian" is in honor of the nineteenth century Norwegian mathematician Niels Abel.) A group of operations is said to be Abelian if it obeys a commutative multiplication law like

$$ab = ba$$

and non-Abelian if it doesn't; that is, if $ab \neq ba$. The multiplication of ordinary numbers is Abelian; the group of rotations of a sphere isn't. Where a sphere will end up after a sequence of rotations depends, in general, on the order in which the sequence is carried out. The group of local transformations of the neutrons and protons (the isotopic spin group), which is also known as the group SU(2) (the special unitary group which can be realized by its actions on the neutron-proton doublet), is non-Abelian. The group SU(1), usually called simply U(1), is an Abelian group and is realized in ordinary local electromagnetic gauge invariance.

In February 1954, Yang gave a lecture on the Yang-Mills theory at the Institute for Advanced Study. From all accounts, it was a disaster. Pauli, who unknown to Yang had been working along related lines, was unmerciful on the question of what the mass of the b-quanta was supposed to be. It was on this point that Pauli had become stuck; it was the reason why he never published his work. When Pauli became fixated on something, wild Brahma bulls were not sufficient to distract him. If the mass were zero, Pauli noted, that would lead to long-range forces, presumably of nuclear strength, among the isotopically "charged" particles such as the nucleons. But no such forces had ever been observed. If, on the other hand, a mass were given to the b-quanta, then, on its face, the gauge symmetry would be destroyed. It was a dilemma to which, Yang and Mills noted in their original paper, "We do not have a satisfactory answer."

It was some 15 years before the dilemma was resolved, but it was only 7 years before Murray Gell-Mann saw how to begin to weave the fragments together into what we now believe is the correct theory of the strong interactions. The inspiration for his synthesis came, as it happened, from developments in weak interaction physics that had occurred since the Glorious Revolution of 1956. It will be recalled that Fermi, when he invented it, modeled the theory of β-decay after quantum electrodynamics as much as possible. Just as in electrodynamics there are currents of charged particles, so in Fermi's original theory there were currents of weakly interacting particles. The fact that the weak interactions were carried by currents — vectors in space and time — had implications for the spectra of the electrons emitted in β-decay. However, it was realized prior to 1956 that currents were not the only possibility for the weak interaction and that the other possibilities made different predictions for the β-decay spectrum. The discovery that parity was violated in β-decay opened up still more possibilities for the interactions. It was a very confusing time because experiments did not seem to lead to a clearcut choice as to which of these several possibilities were correct. Within a year or so, however, some experiments were redone and others done for the first time, and it became clear that Fermi's original guess, with the addition of a second, pseudo-vector, current that was now allowed by parity nonconservation, had been right all along. Furthermore, it was realized that the fact that β-decay was generated by *currents* allowed for the possibility that it might be transmitted by a heavy "photon"—a vector meson. This subject deserves, and will receive, a chapter of its own. Our present concern is the nature of the weak current.

In an important paper published in 1958, Feynman and Gell-Mann argued that the vector part of the β-decay current, as opposed to its pseudo-vector part, should be conserved just as the electromagnetic current is. But, they argued, and, as it turned out, correctly, that the analogy is deeper. The electromagnetic current actually consists of two parts. There is a part that transforms in the isotopic spin space as a singlet like the Λ^0-particle, and there is a part that transforms like a triplet like the three pi-mesons or the three Σ hyperons. In this analogy, the isotopic vector part of the electric current corresponds to the neutral member of the π or Σ triplet that is, the π^0 or the Σ^0. Feynman and Gell-Mann proposed that the charged parts of the isotopic spin current, the ones that correspond to, say, the π^+ and the π^-, are the vector currents that generate β-decay. Actually, that has a number of experimental consequences including what Gell-Mann called weak magnetism, which relates certain β-decay experiments directly to certain aspects of the electromagnetic properties of nuclei. In due course, those experiments were done, and they confirmed this remarkable connection between electromagnetism and the weak interactions.

But that was not all. Gell-Mann realized that the different components of the isotopic spin current obeyed algebraic relations among each other. Indeed, that is just the algebra of the isotopic spin group SU(2). This suggested to Gell-Mann that the road to higher symmetries might be by enlarging the algebra of currents. What Gell-Mann did not realize until the fall of 1960, when the mathematician Richard Block at Cal Tech told him about it, was that there was a name for what he was doing, namely, generating Lie groups, and that it had, since its invention by the Norwegian mathematician Sophus Lie at the end of the nineteenth century, been a well-formulated branch of pure mathematics. Block also told him about the work of the French mathematician Élie Cartan who had, as his doctoral thesis, classified all the simple Lie groups early in the century. This was the work that Gell-Mann was unwittingly proceeding to reinvent.

One could, however, simply look at Cartan's thesis and find the entire catalogue of simple Lie groups and their properties. Most remarkably, the next simplest Lie group after SU(2), namely SU(3), worked. This group contains, as a subgroup, the three currents of SU(2). That must be true if SU(3) is to describe nature, since we know that SU(2) is a symmetry of the strong interactions. But in addition, SU(3) contains five other currents, making eight in all. That is why Gell-Mann called it the eightfold way, taking a text from Buddhist scripture:

> Now this, O monks, is noble truth that leads to the cessation of pain; this is the noble Eightfold Way: namely, right views, right intention, right speech, right action, right living, right effort, right mindfulness, right concentration.

The "cessation of pain" was, in this instance, the stopping of thinking about the wrong theories.

SU(3) made a powerful series of predictions, the most immediately striking of which was that the various particles should be arranged in multiplets. As one might imagine, these multiplets are both more complicated than the isotopic spin multiplets and also contain them. We begin by exhibiting the so-called octet structure of the hyperons of lowest mass (Figure 4-6). It must be understood that, in the limit in which

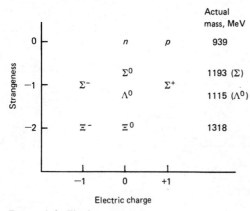

Figure 4-6. The hyperon octet.

SU(3) is meant to be an exact symmetry, these masses are taken to be identical.

In Figure 4-7 we give the corresponding diagram for the family of lowest mass, the spin-0 mesons. When Gell-Mann first discovered this representation, the η^0, which has a mass of 549 MeV, had not yet been found. It was found within the year.

Perhaps the most remarkable representation predicted by the theory was the so-called decuplet, the set of 10 particles shown in Figure 4-8.

It should not be imagined that after Gell-Mann and, independently, the Israeli army officer turned physicist, Yuval Ne'man, proposed this

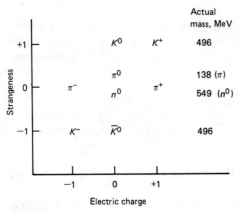

Figure 4-7. The spin-zero meson octet.

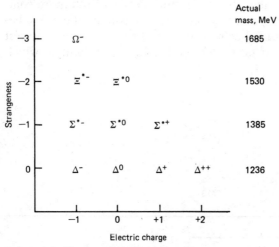

Figure 4-8. The spin-3/2 hyperon decuplet.

scheme, people fell all over themselves to embrace it. Gell-Mann must have had some doubt himself on how it would be received, since the original paper he wrote on it was never published. It was circulated widely as a Cal Tech preprint. Apart from the unfamiliarity of its mathematics, there was the matter of the then unobserved particles, including the Ω^-, for the hypothetical decuplet. Above all, there was the question of the seemingly huge mass splittings. In his unpublished paper, Gell-Mann and then, independently, the Japanese-American physicist Susumu Okubo pointed out that, given simple assumptions about how the SU(3) symmetry was broken, the mass splittings could in a certain sense be predicted. For example, in his paper Gell-Mann noted that the hyperon octet masses would obey a simple "sum rule":

$$\frac{m_n + m_\Xi}{2} = \frac{3m_\Lambda}{4} + \frac{m_\Sigma}{4}$$

The reader may test this equation by using the masses in Figure 4-6. In 1962, at a conference in Geneva, Gell-Mann made an observation about the decuplet masses whose subsequent confirmation made SU(3) believers out of most people.

At that time the Ω^- had not been seen. However, Gell-Mann observed that the mass-splittings in the decuplet should be constant from isospin multiplet to isospin multiplet. Hence, knowing any of the splittings would enable one to predict all the other masses. On that basis he said the Ω^- should be found to have a mass of 1685 MeV. As it happened, at just that time, an 80-inch bubble chamber was under construction at Brookhaven, along with a 5-GeV K-meson beam. One of the physicists working on the bubble chamber, Nicholas Samios, presently the director of Brookhaven, succeeded, with the aid of a note from Gell-Mann, in persuading the then director of Brookhaven, Maurice Goldhaber, to authorize a search for the Ω^- with the new bubble chamber and the new beam. On January 31, 1964, Samios found what he called a gold-plated event: an unambiguous candidate for the Ω^- at just 1685 MeV. SU(3) had come safely home to port.

Gell-Mann's unpublished report on SU(3) began with an example — a hypothetical example — chosen to illustrate how the SU(3) group works. In it Gell-Mann considered a triplet of fictitious leptons which he called ν, e^-, and μ^-. He supposed that the theory, the fictitious theory of these fictitious leptons, was invariant against the transformation of the leptons into each other. The ν, e^-, μ^- triad is the simplest representation of SU(3), and all the properties of the group can be worked out on it. The triad — or triplet — is what mathematicians call the fundamental representation of the group; the n,p doublet is the fundamental representation of the group SU(2).

Gell-Mann used his "leptons" as a sort of scaffold on which to work out the mathematics of SU(3). The saga of how the "leptons" metamorphasized into the quarks has become part of the folklore of elementary-particle physics. Suffice it to say that, in late March of 1963, Gell-Mann visited Columbia University for a week. Over a Chinese lunch early in the week he was asked by Robert Serber, a professor at Columbia, why in his SU(3) theory he did not make use of the fundamental representation: the triplet. Gell-Mann pointed out that if one wanted to use such a triplet to construct a particle like the proton, the building blocks would have to have very peculiar properties. Since the proton has spin ½, the building blocks, which hereafter we shall call quarks, must have a spin of at least ½. The simplest choice is ½.

Now the fun begins. Since the proton has baryon number 1, if we want to construct it out of three quarks, each of the three must have baryon number ⅓. That was the first time a fractional baryon number had been introduced into physics. What about the electric charge? Since the proton has charge 1 and is composed of three quarks, the three must have *fractional* charges, again a completely novel idea in elementary-particle physics. In Figures 4-9 and 4-10 we give the quark and antiquark assignments of baryon number, electric charge, and strangeness.

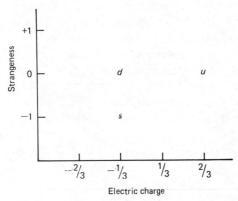

Figure 4-9. The quark table with baryon number ⅓.

The objective, then, is to use the particles shown in Figures 4-9 and 4-10 as the ingredients—the constituents—of the strongly interacting mesons and baryons. We shall not go through the whole list to show that the scheme works—it does—but just give a few representative examples. In considering these examples note that—keeping track of the signs—the charges, strangeness, and baryon numbers of the constituents are added to get those of the constituted particle. Below, we give a representative list.

Spin-0 mesons:

$$\pi^+ \sim \bar{d}u$$

$$K^+ \sim u\bar{s}$$

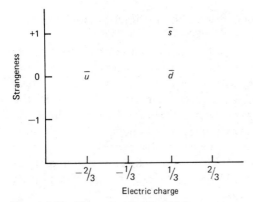

Figure 4-10. The antiquark table with baryon number ⅓.

Spin-1 mesons:

$$\rho^+ \sim \bar{d}u$$

$$K^{*+} \sim u\bar{s}$$

Spin-½ baryons:

$$p \sim uud$$

$$\Sigma^+ \sim uus$$

Spin-3/2 baryons:

$$\Xi^- \sim dss$$

$$\Omega^- \sim sss$$

A few remarks are in order. When two spin-½ objects are combined, the result can have spin 1 or spin 0. Thus the quark model predicts, and experiments confirm, that spin-1, strongly interacting mesons like the ρ and the ω^0 should form an SU(3) octet. In Gell-Mann's original paper, and for a few years afterward, it was put forward that these vector mesons were the Yang-Mills b-quanta for the strong interactions. As we shall see, that notion turned out to be a red herring. This led to all sorts of confusion about vector meson mass, which turned out, once the true picture emerged, to have been irrelevant. The second point to make is that if we give the s-quark a mass slightly different from the masses of the d- and u-quarks, we can see why the equal spacing rule for the decimet is true. To go from one row of the decimet to the next, one simply replaces a d-quark by an s-quark.

Finally, and remarkably, the whole scheme was discovered in two places simultaneously and independently. A postdoctoral student from Cal Tech named George Zweig, who was working at the CERN laboratory in Geneva, developed the whole affair for himself; he called the fundamental triplets aces. His paper was never published, and the name "quark" stuck. The scheme, by any name, is very elegant, but if it is accepted, we come back to Mach's question, "Do quarks exist?" or, in other words, "Have you seen one?" That is the subject of the next chapter. In 1969 Murray Gell-Mann was awarded the Nobel prize in physics for, among many other things, his work on SU(3).

5
The Bound
and the Free

Ernest Rutherford, in the laboratory, was affectionately referred to by his associates as a "thundering nuisance." He roared and thundered in the vicinity of delicate equipment while spewing cigarette ashes in all directions. When things were going well, he sang, if that was the word for it, a tuneless version of "Onward Christian Soldiers." He also happened to be one of the greatest experimental physicists who ever lived. His intuition as to how physical phenomena worked, or ought to work, was remarkable. Late in his life—he died in 1937 at the age of 66—he was collaborating with a research team at Cambridge headed by Mark Oliphant that was using deuterium as a projectile in collisions by which they hoped to produce tritium—superheavy hydrogen. The results were baffling. No one could figure them out. Oliphant describes the denoument.

> Rutherford produced hypothesis after hypothesis, going back to the records again and again and doing abortive arithmetic throughout the afternoon. Finally we gave up and went home to think about it. I went all over the afternoon work again, telephoned Cockroft [John Cockroft] who had no new ideas to offer, and went to bed tired out. At three o'clock [A.M.] the telephone rang...my wife...came back to tell me that "the Professor" wanted to speak to me. Still drugged with sleep I heard an apologetic voice express sorrow for waking me, then excitedly say "I've got it. Those short-ranged particles are helium of mass 3" [a heretofore unobserved light isotope of helium]. Shocked into attention, I asked on what possible grounds could he conclude that this was so, as no possible combination of twice two [deuterons hitting deuterons] could give

two particles of mass 3 and one of mass unity. Rutherford roared, "Reasons! Reasons! I feel it in my water!" And he was right.

In 1909 Rutherford, who had been born in New Zealand, was at Manchester University in England. He had a young student named Ernest Marsden who was in need of an experimental problem to work on. Rutherford possessed a few milligrams of radium, which is an alpha-particle emitter. He used a thin pencil of the energetic alpha particles as a beam to scatter from a target, a foil of aluminium or gold. With this setup he and Hans Geiger — of the counter — who also was at Manchester, had succeeded in measuring the electric charge of the alpha-particle and so had made the identification of the α-particle with the helium nucleus possible. Now Rutherford proposed to Geiger that he and Marsden use the same setup to see if, when scattered from, say, gold, any of the α-particles would be scattered through a large angle.

Why Rutherford wanted that experiment done is not clear. He must have had some intuition which he himself may not have been able to formulate. He expected that there would be no scattering at large angles; that essentially all of the α-particles would pass through the gold foil like a bullet through smoke. That is not what happened. A few scattered by more than 90°; they bounced back in the general direction from whence they had come. In reminiscing about it, Rutherford recalled that "It was quite the most incredible event that has ever happened to me in my life. It was almost as incredible as if you fired a 15-inch shell at a piece of tissue paper and it came back and hit you." The atomic nucleus had been discovered.

One may speculate about what our attitude toward the "existence" of the nucleus might have been if the only evidence for it were the sort of scattering experiments begun by Rutherford in which the nuclear constitutents are not observed. In fact, we can take nuclei apart by, for example, bombarding them with gamma-rays, the process of photodisintegration discovered by James Chadwick and Maurice Goldhaber in 1934. We can "see" the constitutent parts — the protons and neutrons — emerge, which gives us a feeling of confidence when we say that an atomic nucleus is made up of protons and neutrons. Now what of the quark? In what sense is it a constituent of a particle like the proton? When the quark was first invented by Gell-Mann and Zweig, there were two attitudes toward that question. Quarks might be "real" particles, in which case we would expect them to be directly observable like the neutrons and protons in the nucleus, or they might be "purely mathematical entities" (Gell-Mann's phrase), in which case we would never observe them directly. That would give new meaning to the notion that the elementary particles are "made out of" quarks. The situation that would be most familiar to us would be the former. We have been there before. Matter is made up of atoms; atoms are made up of nuclei and electrons; and nuclei are made up of protons and neutrons. Indeed, no sooner was the quark hypothesis made known than people began looking for the quark. They are still looking. They have looked in the sea, on the moon, and in laboratory experiments of various kinds. To this day, no one has seen a free quark. Why then do we believe the quark exists? To answer that question is the burden of the rest of this chapter.

It might be satisfying to report that some historically minded experimental physicist, having read Rutherford's account of his discovery of the atomic nucleus, would have consciously set out to duplicate it by scattering high energy projectiles from, say, the proton with the goal of observing anomolous large-angle scatterings. That is not how it happened. In 1966, the 2-mile-long Stanford linear accelerator, with the capacity of accelerating electrons up to 20 GeV, was completed. Quarks were certainly not a motivation for constructing it. It was clear that the accelerator could be used to extend the kind of elastic scattering experiments that Hofstadter and his group had been doing and that it could also be used to test quantum electrodynamics at high energies. Beyond that, it had been built for the best of reasons: No one knew what new physics it might find. No one ever really knows. In the beginning, what was found with it was not very spectacular. The elastic scattering results joined smoothly to Hofstadter's, and quantum electrodynamics remained valid. But, in 1967, an MIT-SLAC (for Stanford linear accelerator) group decided, just to see what would happen, to measure inelastic electron-proton scattering. "Elastic scattering" means that the electrons and protons collide and, like billiard balls, retain their identity. "Inelastic scattering" means that, in the collision, the proton can break up in various ways.

To understand what the group found, it is important to understand a little about how the experiment works. The electrons hit the proton target and recoil with different energies, depending on what has happened to the proton. The energies can be sorted out by using an electron spectrometer, the same kind of device that is used to measure the electron energy spectrum in β-decay. What happens is that the spectrometer is set up at a given angle to the beam, and then the relative number of electrons at various energies which have recoiled at that angle is measured. When that has been done, the spectrometer can be moved to a different angle and the work begun again. In these experiments the debris emanating from the proton is not observed. The group had an idea of what to expect. All the recoils from the elastic scattering, at a given angle, have about the same energy. Hence, the group expected to see an "elastic peak" centered at that energy. But the proton can be excited by the scattering; it can turn into a Δ^+, for example. If it gets excited into one of these resonant states, that also will show up as a peak in the plot. Beyond those resonances, the group expected the plot to fall off rapidly. They were not anticipating any hard structure in the proton, so there was no reason, they thought, to expect anything else.

The question that might well be asked, given that Gell-Mann and Zweig had argued 3 years earlier that the proton should consist of three quarks, is why the experimenters didn't expect to find something like what Rutherford had found for the atomic nucleus. Part of the reason, perhaps most of it, is that very few people took the quark model entirely seriously. The dynamics were completely mysterious. What held the quarks together? Did they, in turn, have any structure? By comparison, one should keep in mind that Rutherford understood very well the dynamics of the α-particle scattering. Given that the α-particles were scattering from a point, electrically charged nucleus, he could compute what happened in detail. He could compute just how many particles scattered at a given angle, and the comparison with experiment is what gave credence to the model. On the other hand, at the time of the first SLAC experiments, no one, even if someone had wanted to, had the foggiest notion of how to do the same thing for the quarks.

In any event, by 1968, it was clear that the SLAC data were very odd. Figure 5-1 is a schematic graph of the electron energy spectrum plotted at some characteristic angle. The fact that essentially the same plot emerges at different angles is what builds confidence that something real is being observed. The unexpected feature of the graph in Figure

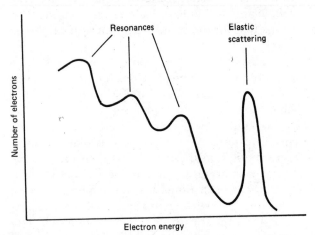

Figure 5-1. The energy spectrum of the electrons observed in elastic scattering experiments done at SLAC.

5-1 was that, for small energies corresponding to collisions in which the electron gave up the most energy, the curve was much higher than anyone had predicted, at least 30 times higher. A few brave souls like Wolfgang Panofsky, the director of SLAC, speculated publicly that this shoulder might mean that the electrons were bouncing off something hard in the proton. Indeed, James Bjorken, also of SLAC, developed a theory, which later turned out to be essentially correct, in which those objects were quark-like. But not much attention was paid. Then came Feynman. In August 1968, he got his first look at the SLAC data. When plotted in a certain way, it exhibited what Bjorken had called scaling.

In principle, the functions that describe the inelastic scattering process can, and will, depend on the rest mass of the target particle, in this case the proton. They might possibly depend on the rest mass of the projectile, but in these experiments the projectiles are electrons and neutrinos, and they have a negligible mass compared to the kinetic energies involved. (Inelastic neutrino scattering analogous to the electron scattering done at SLAC began in 1971, at CERN, when the gigantic 197-inch bubble chamber Gargamelle began registering events.) It is not clear that the 1-GeV mass of the target protons can be neglected, but Bjorken, in an inspired guess, did so anyway. This meant that the functions depended on a single variable. That is what the data seemed to show, but what did it mean?

To understand those data, Feynman invented what became known as the parton model. He supposed that the proton was composed of spin-½ objects: "partons." In his model these objects were not necessarily to be identified with the quarks; the quarks are a special case in which the partons also are a fundamental representation of SU(3). The latter we can call the quark-parton model, and it, as we shall see, makes additional predictions. The idea is that, in the scattering, each parton contributes "incoherently" to the cross section, which means that, to find their total effect, we simply add the individual contributions.

Feynman could not really justify that assumption, since he did not try to specify the dynamics of the partons. Each of the partons carries a fraction of the proton's momentum, and what Feynman realized was that the function that was being measured in the inelastic electron-proton scattering could be interpreted as the function that described how that fraction of the momentum was distributed. Bjorken scaling emerged as a deduction from the model, with the extra bonus that it provided a simple way to visualize what it meant. If the assumption that these partons are quarks also was made, much stronger predictions were possible.

For example, if the same experiments are done for neutrons and the results are compared to proton scattering, the quark-parton model predicts that the ratio of the structure functions for neutron versus proton scattering should, over its range of arguments, always be less than or equal to 4 and greater than or equal to ¼. That is borne out by experiment. There are serious complications, however. If the model is applied naively, it predicts that this ratio should be exactly 2:3. That is not borne out by experiment except for large values of the parameter that measures the fraction of the proton's momentum taken up by an individual quark. For small values of the parameter the ratio approaches unity. This departure from the naive quark model is interpreted as reflecting effects of the quark binding. We shall come back to these matters when we discuss the dynamics of the quarks.

While all this activity was going on at Stanford and CERN, theorists had become aware that the constituent quark model of elementary particles was in serious trouble. This had to do with the Pauli exclusion principle. We can see what was involved most clearly by examining the quark content of the Ω^-. The Ω^-, it will be recalled, is a spin-3/2 particle with a strangeness of −3. In the quark model the Ω^- is taken to be a bound state of three s-quarks. The most natural bound state of three spin-½ objects which produces a particle of spin 3/2 is simply to put all three in a common state with no orbital angular momentum. That is generally the most tightly bound configuration, since the addition of an orbital angular momentum tends to make the particles avoid each other. But the Pauli principle forbids such a state, since identical spin-½ objects cannot be in the same state. If one insisted on maintaining the Pauli principle for the quarks, the only way to construct the Ω^- would be by some very unnatural gyrations with angular momentum.

The first suggestion that the way out of that dilemma was to give up the Pauli principle for quarks—by introducing statistics, allowed by quantum mechanics, that were neither Bose-Einstein nor Fermi-Dirac—was made in 1964 by Oscar Greenberg. He called these intermediate statistics parastatistics. It is probably fair to say that, since most physicists were not taking the quark model entirely seriously at that time, this complication was regarded as more of an obstacle than an opportunity.

In 1965, in an extraordinarily prescient paper, Moo Young Han and Yochiro Nambu introduced a set of ideas which, when they were developed some years later, became the modern theory of the strong interactions. In the first place, they noted that the Pauli problem could be solved if there was more than one triplet of quarks. Their original idea was to get rid of the Pauli problem and fractionally charged quarks in one blow. All the Han-Nambu quarks—there are nine arranged in three triplets—have integral charges. To construct a particle like the Ω^-, one takes a quark from each of the different triplets with charges that add up to −1. Since the quarks are distinct, the Pauli problem is solved.

But there is another possibility. One can take three triplets with the same set of fractional electric charges as the original Gell-Mann-Zweig quarks but distinguished by an extra quantum number which is now universally called color. "Color" has nothing to do with pigmentation; it is just a mnemonic label for the distinct quarks. The reader can choose any set of three colors which he or she finds aesthetically pleasing, although it is useful to reserve white for the physically observed particles — in this instance, white means the absence of color. This second scheme, it turns out, is mathematically equivalent to Greenberg's parastatistics. Although they did not work out the details, Han and Nambu went on in their paper, almost casually, to set out what became the program for constructing the strong interactions, something that Gell-Mann would later name "quantum chromodynamics."

To visualize what Han and Nambu proposed, let us lay the three quarks, with their colored clones, out in a little array, where the subscripts r, g, and b refer to red, green, and blue.

	Red	Green	Blue
Up	u_r	u_g	u_b
Down	d_r	d_g	d_b
Strange	s_r	s_g	s_b

If we make all the quarks massless, we have the possibility of two distinct SU(3) groups. By interchanging the quarks vertically, we have the original SU(3) model of Gell-Mann and Ne'eman. We may call this the SU(3) of "flavor," the flavors here being up, down, and strange. By interchanging the quarks horizontally, we have an entirely different SU(3) group, which we may call the SU(3) of color. Here the colors are chosen arbitrarily to be red, green, and blue. Han and Nambu made two extremely important suggestions concerning this second group $S(U)_C$, where C is for color. The first suggestion was that all observed strongly interacting particles should be in the singlet representation of $SU(3)_C$. A less technical way to put that is to say that all strongly interacting particles observed should be colorless. That immediately puts an enormous constraint on the quark dynamics. It should be of such a character that no one will ever see a free quark. To repeat, quarks are, according to this idea, permanently confined inside such particles as the neutron, the Λ^0, and the Ω^-.

Han and Nambu went on to conjecture what this dynamics might be. They proposed that, in the spirit of Yang and Mills, the $SU(3)_C$ symmetry be made local. This immediately brings into being eight Yang-Mills gauge particles, the old b-quanta particles or colored "photons," which came to be called gluons, the "glue" that holds the quarks together. Since $SU(3)_C$ is supposed, in this picture, to be an exact symmetry, these gluons must be strictly massless; and since they carry color, it is a prediction of the scheme that, like the quarks, they too will never be observed as free particles. These ideas, which do not seem to have been worked out in greater detail by Han and Nambu, remained buried in the literature for the next 8 years.

In the meantime, an entirely new accelerator technology had been developed. The merit of this new technology can be understood in the commonsense observation that it is the head-on collisions that do the most damage. Figure 5-2a shows a collision between two particles of equal mass, say, protons or electrons, as viewed in what is called the center-of-mass or, equivalently, the barycentric system. Figure 5-2b shows the collision as viewed in a system in which one of the particles is at rest. As a rule, the Figure 5-2b system is called the laboratory system

Figure 5-2. A collision between two particles (a) as viewed in the center of mass and (b) as viewed in the laboratory.

because, in the collisions we have been considering so far, the target (the bubble chamber, for example) is at rest in the laboratory.

Because the head-on collisions are the most probing, experimenters use as a benchmark the amount of energy that would be available in the center of mass for a given collision. It is not difficult to show that if all the kinetic energies are much larger than the rest masses of the colliding particles, the center-of-mass energy increases only as the square root of the laboratory energy increases. That means, for example, if we want to make 10 times more energy available in the center of mass, we must raise the energy of the conventional fixed target accelerator by a factor of 100.

On its face, that is a very discouraging piece of news. But as is traditional in the accelerator business, someone came along with a better idea. In this case the innovator was Gerard O'Neill from Princeton who, in 1957, while on leave of absence at Stanford, proposed the notion of the colliding-beam accelerator. The first version, which was completed at Stanford in 1962, consisted of two separate rings each about 10 feet in diameter. The rings touched, making the whole configuration look like a figure 8. Each ring contained electrons which circulated in opposite directions and could collide at the place where the two rings joined. The electrons in each ring had energies of up to about 300 MeV, which meant that a center-of-mass energy of about 600 MeV was available. Since the collisions between the two beams was head-on, all of this energy was available to make new particles in those collisions.

At about the same time, an Italian group led by Bruno Touschek began developing a machine that would have only one ring. It held both electrons and positrons circulating in opposite directions in orbits that could be made to intersect occasionally. The prototype, which began operating in 1961, had storage rings that were about 3 feet in diameter. By 1964, the Italian group was reporting results from collisions of 200-MeV electrons with 200-MeV positrons, and the era of electron-positron colliders had begun. By the early 1970s there was an electron-positron collider, the CEA, with a net energy of about 5 GeV in Cambridge, Massachusetts, a more advanced collider, the SPEAR, with a net energy of about 8 GeV at Stanford, and a larger version of the Italian machine, the ADONE, with a net energy of about 3 GeV. To understand what the experimenters at those machines expected to see, consider the Feynman diagram in Figure 5-3. In the diagram the elec-

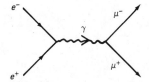

Figure 5-3. A Feynman diagram showing the process $e^+ + e^- \rightarrow \mu^+ + \mu^-$.

tron and positron annihilate into a single photon. That, in turn, creates a lepton pair which, for the sake of definiteness, we have made a muon and antimuon. This is the process that most physicists expected the machines to produce most of the time. A few venturesome souls, notably Bjorken, expected the diagram in Figure 5-4, where q and $\bar{q}$ are a quark-antiquark pair, to be of comparable importance. Since free quarks were not expected to be observed, the quark-antiquark pair reconstitute themselves into a shower of hadrons as indicated in Figure 5-4. In the quark-parton model the production of each quark type is taken to be an independent process. The electromagnetic interaction of the charged quarks with the photon is much like that of the photon with the leptons. The difference is the electric charges, which are fractional

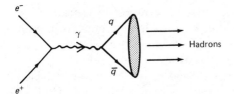

Figure 5-4. A Feynman diagram showing the process $e^+ + e^- \rightarrow$ hadrons.

for the Gell-Mann–Zweig quarks. If the cross section for hadron production is plotted against the cross section for the production of muons, this ratio, conventionally called R, is according to this picture given simply by

$$R = \Sigma_{\text{flavor}} \, Q_i^2$$

where Q_i is the charge of the ith flavored quark. For the original quark model then,

$$R = \frac{4}{9} + \frac{1}{9} + \frac{1}{9} = \frac{2}{3}$$

If color was included, the ratio became

$$R = 3\left(\frac{4}{9} + \frac{1}{9} + \frac{1}{9}\right) = 2$$

By 1973 there were two measurements of R, one from ADONE in an energy range from 1 to 3 GeV which showed R to be about 2, just as the quark-parton model with colored quarks predicted, and another from the CEA in Cambridge. This was at an energy of between 4 and 5 GeV, and it produced the very mysterious result that R appeared to have jumped to over 4 — something that provoked disbelief. As it turned out, it was a portent of things to come.

In the meantime, the theorists had not been idle. Indeed, 1973 was a banner year. Various groups of theorists began working on the quantum chromodynamical model of quark binding which Han and Nambu had sketched out 8 years earlier. Very quickly a discovery of capital importance was made by David Gross and Frank Wilczek at Princeton and, independently, by David Politzer at Harvard. Their discovery became known as asymptotic freedom. It had its roots in some seminal work done by Gell-Mann and Francis Low in 1954. Gell-Mann and Low asked themselves, what is the force between two electrons at very short distances? To understand the question, let us exhibit the diagram that in

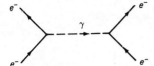

Figure 5-5. A diagram of one-photon exchange between two electrons.

quantum electrodynamics leads to Coulomb's law (Figure 5-5). However, at short distances, which correspond to large momentum transfers between the electrons, radiative corrections play a dominant role. The simplest such diagram, known as a vacuum polarization diagram, is given in Figure 5-6. What Gell-Mann and Low showed is that, quite generally, the effect of adding all these diagrams is to introduce a space-dependent electric charge $e^2(r)$ so that the corrected Coulomb law takes the form $e^2(r)/r$. Moreover, and this was the crucial point, this charge gets *larger* as the distance between the two electrons decreases. Therefore, the force between the particles gets stronger the closer the particles are together over and above the increase due to the $1/r$ dependence of the Coulomb force. When the particles are at macroscopic distances from each other, the function $e^2(r)$ approaches a constant. Indeed, this is just the constant we identify with the observed electron charge $e^2 \simeq \frac{1}{137}$.

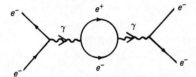

Figure 5-6. The lowest-order vacuum polarization diagram.

A folkloric argument that went along with this was meant to explain why the charge at large distances was smaller than the charge at short distances. The reason, the argument ran, was that the positrons from the vacuum shield the electron's bare charge and therefore the electron's observable charge is smaller. The trouble with the argument was that it was too persuasive. It convinced everyone, for nearly 20 years, that it was a general truth about all quantum field theories. Then came the calculations of 1973. Those calculations were made, for the first time, for a Yang-Mills theory. That theory has, as we have noted, the distinctive property that its "photons," that is, gluons, for quantum

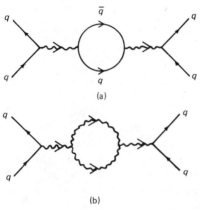

(a)

(b)

Figure 5-7. (a) Quark-gluon and (b) gluon-gluon
vacuum polarization diagrams.

chromodynamics couple directly to each other. Thus, in addition to
quark-gluon diagrams of the kind shown in Figure 5-7, which produce
the Gell-Mann–Low behavior, there is another class of diagrams, of
which samples are given in Figure 5-7, that are completely novel.

Diagrams of the second class produce what we might call the
anti–Gell-Mann–Low behavior. If we call $g(r)^2$ the quark-nucleon coupling constant, which because of the structure of the Yang-Mills theory
is in fact identical with the quark-quark coupling constant, these new
diagrams give $g(r)^2$ the property that, as r becomes large, $g(r)^2$ approaches infinity and, as r approaches zero, $g(r)^2$ approaches zero. Before commenting on the physical significance of this remarkable result,
we should note that the actual behavior of $g(r)^2$ is determined by adding
up both classes of diagrams. Hence, if we are going to get anti–Gell-
Mann–Low behavior—that is, asymptotic freedom—the contributions
from the first class of graphs must not overwhelm those from the second class of graphs. By making certain plausible assumptions, this can
be translated into a limit on the number of quark flavors allowed by the
requirement that the theory by asymptotically free. This number turns
out to be less than 16. We shall have more to say about quark flavors
shortly; now to the physics.

The apparently ineluctable paradox about the quark-parton model is
that two things that appear to be mutually incompatible are demanded
of it. On the one hand, the bound quarks are free—they behave like
free particles inside the hadrons—and on the other hand, the free
quarks are bound—they are permanently trapped inside the hadrons.
The asymptotically free quantum chromodynamics seems to have just
this property. For example, when a quark and an antiquark are in the

vicinity of one another, the effective coupling constant becomes very small. The two particles are essentially free. But when they are widely separated, the constant becomes very large and it becomes impossible to separate them. This is the quantum chromodynamic picture of a meson; it is the qualitative picture. Making it quantitative has absorbed, and is still absorbing, a small army of hard-working theorists.

One aspect of the work is of particular interest. Field theories have traditionally been formulated in a space-time continuum, which means that the variables at which the fields are evaluated can run continuously over all the real numbers. From time to time, various theorists have suggested that the use of a space-time continuum might be unphysical, since it involved distance intervals that were arbitrarily small and hence beyond the possibilities of measurement. Some attempts were made to introduce a "fundamental length" which would make space-time discrete and thus have the effect of replacing the differential equations of quantum mechanics by difference equations—sums—of the sort that are used by a computer when it solves such equations. When the problem of the infinities in quantum field theory came under study after World War II, there were suggestions that this fundamental length might serve as a "cutoff" which would render the divergent quantities in field theory finite. That never really attracted a great deal of favorable attention because it was much simpler to work with the continuous quantities and, with renormalization, to extract finite answers.

With the advent of quantum chromodynamics, which is an extremely difficult theory to calculate with, and the advent of high-speed computers, interest in the discrete theories was renewed. These go under the rubric "lattice gauge theories," since the space-time points are taken to be the coordinates of a lattice—a grill—with a spacing that acts like a fundamental length. One can, and many theorists do, take the attitude that this discretization of space-time is simply an artifact that enables one to work with finite sums and that after a discrete computation is completed, one then passes back to the continuous limit. A few theorists, on the other hand, believe that space and time are *really* discrete and that physics at very small distances may really be some sort of lattice physics.

In either case, the procedure is the same, namely, to put the difference equations on a computer. In fact, a new university cottage industry has developed to build special computers that are dedicated to the one task of computing quantities in quantum chromodynamics. These "QCD machines" are, because of their specialized design, as powerful as the most sophisticated supercomputers, although they lack the flexibility. The results that have so far emerged look encouraging. Some of the work has to do with whether the theory really confines quarks. Although that has not yet been proved, all of the computer work so far is consistent with confinement. There has also been work on trying to compute the mass spectrum to see if it can be understood why the hadrons have the distributions of masses they have. This work also looks promising. A nice feature of the computer work is that it is relatively inexpensive. Both the engineering and the construction are done by the physicists themselves with parts that can be ordered directly from a catalog. It may be the closest thing to sealing wax and string that we have in high-energy physics.

In 1970, Sheldon Glashow, John Iliopoulos, and Luciano Maiani published a paper entitled "Weak Interactions with Lepton-Hadron Symmetry." Their paper turned out to be extremely important, although when it was published, it seemed very speculative. The authors were concerned about why certain weak processes which seemed to be allowed by the Fermi theory, at least in higher order, did not seem to happen at all—shades of the μ-e-γ problem. An example of such a decay process is $K^0 \rightarrow \mu^+ + \mu^-$. This decay, which has never been observed, has an upper limit, relative to the principal decay modes, of less than 1 in 100 million. In the conventional models, as it happens, this decay was forbidden in lowest order but was allowed as a radiative correction. Since the theory of weak interactions, then in use, was not renormalizable, many theorists took the attitude that this higher-order effect was not something to worry about and would somehow disappear when a renormalizable theory was invented.

On the other hand, Glashow and company took the position, correctly as it turned out, that this was a serious problem and should not be swept under the rug. However, they proposed a solution which, at first sight, seemed to many people to be worse than the disease. They proposed to introduce yet a new quark which, following a terminology of Bjorken and Glashow, they called the charmed quark. They noted that this quark, if it were introduced in just the right way, would have the effect of canceling the other diagrams contributing to, for example, the decay $K^0 \rightarrow \mu^+ + \mu^-$. (That came to be called, after the authors, the GIM mechanism.) This seemed almost too good to be true, and since it involved an entirely new quark, it was at first not greeted with a great deal of enthusiasm. Then came the November Revolution of 1974. When we left it in 1973, the R ratio—the ratio of hadronic to leptonic events in e^+e^- annihilation—had risen to over 4 in the region of about 4 GeV. This result, which was first found at the CEA accelerator in Cambridge, was then confirmed at SLAC. We argued that with three flavors of quarks, u, d, and s, and three colors, R would have the value 2. If we add the charmed quark, which has a charge $2/3$ and assume it too comes in three colors, R is increased to 3.33. This was in the right direction to fit the data, but the data kept on increasing.

What happened next is one of the sagas of postwar elementary-particle physics. Let us start with the west coast. In early November 1974, it became clear to the SLAC-Berkeley group that they were seeing something very peculiar in the hadron production in e^+e^- annihilation in the region of 3.1 GeV: a striking rise in the production rate. This was an energy regime that the group had already scanned twice, and there was considerable pressure to go on to higher energies. But by Friday, November 8, they decided to back down over the 3-GeV regime one more time. By Sunday morning it had become clear that they had made one of the most extraordinary discoveries in elementary-particle physics of the decade. At exactly 3.105 GeV the counting rate had jumped by a factor of 100. At a few MeV on either side it had dropped down to normal. A new resonance had been found, and it turned out to have a mass of 3096.9 MeV with a width of only 0.06 MeV—a needle in a haystack. Such a width was completely out of line with any of the other observed resonances like those of the ρ and the ω^0, which have widths of something like 100 MeV. Such a small width corresponds to a lifetime of 10^{-2} second, extremely long for a strong interaction resonance.

To compound matters, on November 21 a second such peak was discovered, this one at 3685 MeV with a width of 122 MeV. Now to the east coast. In 1972, Samuel Ting of M.I.T. had gotten the notion to study the production of electron-positron pairs produced in the collisions between protons and nuclei. That is almost the inverse of the SLAC experiment. Also, it is one that was extremely difficult, since these leptons would be drowning in a background of hadrons. By 1974, Ting had a group at Brookhaven prepared to take data on the lepton pairs. By September, the group was beginning to find evidence that in the region of 3.1 MeV—yes, the same 3.1 MeV being examined at SLAC—there was a clustering of events indicating that the e^+e^- pairs were coming from the decay of a particle with a mass of about 3.1 MeV. As the group prepared to publish these results, they were completely unaware of what was happening at SLAC. But the very Sunday on which the SLAC group had found their resonance, Ting, by coincidence, had gone to Stanford to attend a committee meeting at SLAC. The following morning two groups were able to compare their data and to conclude that they had found the same particle. It is now called the J/ψ (3097), and the heavier particle found at SLAC is called the ψ (3685). But what are they?

Actually, some theorists had anticipated the right answer even before the experiments were made. People who had become familiar with the

QCD way of looking at the structure of mesons conjectured that if it were really true that there was an additional quark-c — for "charmed" — it should bind with its antiquark-$\bar{c}$ to produce charmless mesons — hidden charm — and also with one of the other quarks to produce charmed mesons — naked charm. That is, there should, along with $c\bar{c}$ states, be $c\bar{u}$, $c\bar{d}$, and $c\bar{s}$ states. The original $c\bar{c}$ state, the J/ψ (3097), has orbital momentum 0 and spin angular momentum 1. That is, the spins of the c and $\bar{c}$ line up to form a state of total spin angular momentum 1. The ψ (3685) has the same configuration of angular momenta, but there the $c\bar{c}$ are in an excited state in analogy to the comparable excited state of positronium. Just as it happens in positronium, one would expect to find, among the charmonium states, a state of zero orbital angular momentum and zero spin. Indeed there is such a state, the n_c, with a mass of 2981 MeV. By now an entire small universe of charmonium states has been identified, a beautiful little laboratory for testing the ideas of quantum chromodynamics.

As we have mentioned, in order for the charmonium interpretation of the new resonances to hold up, one must also have evidence for naked charm. That is, there must be particles that explicitly exhibit a charm quantum number, an analogy to strangeness. In these mesons there is a single charmed quark bound to either a u-, d-, or s-quark. The decay of such a meson was predicted to take place when a charmed quark converted into a strange quark by a weak interaction. Hence, the hallmark of a meson with naked charm is that the meson should be a narrow resonance decaying predominantly into K-mesons and leptons. In 1976 the first such meson, the so-called D^0, was definitively identified at SLAC. A short time later the positive counterparts were discovered, all of which convinced even the skeptics that charm was here to stay.

But that was not all. In 1977 a group at the Fermilab announced a sharp peak in the distribution of $\mu^+\mu^-$ pairs coming from the scattering of protons on heavy targets. The peak was at 9460 MeV and had the incredibly narrow width of 0.04 MeV. It was, needless to say, interpreted as the bound state of yet a new quark-antiquark pair. This quark has been given the somewhat inelegant name bottom, or b. There are reasons to believe it may have an analog, the t, or top, quark, although the corresponding particles have not yet been found. Bound states of the class of $b\bar{b}$ are called Υ (upsilon) particles, and several have been identified. There are also — forgive the expression — naked bottom mesons, the family of B-mesons. These have recently been observed, and they decay prominantly into D-mesons.

Where will it all end? How many flavors are there? When we discuss the interplay of cosmology and elementary-particle physics, we will review the arguments that indicate that, if we include the top quark, we may have seen all the flavors we are going to see. If that is true, it is a deep mystery, since we have just begun to explore the scale of energies that are available to us in principle. Why are there so few quark flavors? No one knows.

What then about Mach's question: Do quarks exist? What would be the response to the question, "Have you seen one?" Just how tantalizing that question is can be illustrated by the matter of the "jets." The picture of an electron-proton scattering event that emerges from the quark-proton model is that an electron interacts with a single quark in the proton while the remaining quarks stand by as spectators. The reacting quark should recoil from the collision and carry the electron's momentum. Since, according to the theory, we can never observe a naked quark, the effect should be to produce a collimated beam of hadrons, a jet produced by the escaping quark in the process of clothing itself.

Moreover, in an e^+e^- collision which produced a highly energetic $q\bar{q}$ pair moving in opposite directions, the argument went, there would be two jets emitted back to back. Furthermore, one could imagine events in which one of the quarks emitted a gluon which, since it also is unobservable as a naked particle, would produce an additional jet, a trijet event. In fact, all three of these phenomena have been observed and studied in detail. That, and the rest of the evidence we have presented, leads to the conclusion that even if quarks don't exist, they would have to be invented.

6
The Broken
and the Unbroken

In the preceding chapters we have seen examples of a remarkable scientific phenomena: discoveries without context. These are extremely important discoveries, usually theoretical, which go unappreciated for long periods of time because there is no context into which they fit. When the context is finally created, the discoveries find their natural places. An example was the discovery by Yang, Mills, and Shaw, in 1954, of the non-Abelian gauge theories, which did not find their places in elementary-particle physics for nearly 20 years. Another example, which will be relevant when we take up the subject of the next chapter, is cosmology.

In the late 1940s George Gamow and his two young associates, Ralph Alpher and Robert Herman, published a series of papers which, collectively, made the prediction that the present universe should be filled with a uniform background of microwave radiation with a temperature of about 5 kelvins, radiation left over from the Big Bang. The arguments that led to that prediction, which we will outline, would in their essence stand up in a modern classroom. They begin with the recognition that the first step required in forming the light elements from free protons and neutrons in the early universe was the formation of deuterium, heavy hydrogen with a nucleus consisting of one neutron and one proton. This formation takes place by the process of radiative capture; that is, $n + p \rightarrow d + \gamma$. The process begins soon after the Big Bang but it is not effective in producing substantive amounts of deuterium at those enormous temperatures because the inverse process $\gamma + d \rightarrow n + p$ destroys the newly formed deuterium. In this epoch, deuterium is like early fall snow which melts as soon as it hits the ground.

However, the universe cools as it expands. It has been known since the mid-1930s that if the expansion of the universe conserves entropy, then the scale factor, usually called R, must increase as the inverse of the temperature; that is, $T \sim 1/R$. It has also been known since the 1930s — this involves a mixture of general relativity and kinetic theory — that, in the early universe, the rate at which the universe was expanding varied with the square of its temperature. Therefore, the expansion rate was decreasing rapidly as the universe cooled. When the temperature reached something like 10^{10} kelvins, the rate of deuterium formation became comparable to the expansion rate, and when the temperature dropped by another factor of 10 to about 0.6×10^9 kelvins, the ambient photons no longer had enough energy to break the deuteron apart and the production of deuterium took place rapidly.

Gamow and his collaborators used that condition to determine the density of neutrons and protons in the universe at that temperature. They found it to be about 10^{18} nucleons per cubic centimeter, or about a million times less dense than ordinary matter. That corresponds to a mass density of about 10^{-6} gram per cubic centimeter. Suppose we call this density n_B, where B stands for baryon. As the universe expands, n_B decreases; it falls off inversely with the volume, V:

$$n_B \sim \frac{1}{V} \sim \frac{1}{R^3} \sim T^3$$

Therefore, as Gamow and his collaborators realized, if we take the ratio of the baryon density at the time of deuteron formation to the present baryon density, which we may call n_B^0, and use the relation between T and R, we have the following proportionality:

$$\frac{n_B}{n_B^0} = \left(\frac{T}{T^0}\right)^3$$

Hence, if we know the present baryon mass density, we can compute the present temperature. That density can be found by direct measurement. The number Gamow and his collaborators used was about 10^{-29} gram per cubic centimeter. Hence we have the following equation for the present temperature:

$$\frac{10^{-6}}{10^{-29}} = \left(\frac{0.6 \times 10^9}{T^0}\right)^3 \quad \text{or} \quad T^0 \simeq 5 \text{ K}$$

This result was published by Alpher and Herman in 1949 in the *Physical Review*, the least obscure physics journal. But the relic Big Bang radiation, which anyone could have been discovered by looking for it anytime after 1949, was not actually discovered until 1964, by Arno Penzias and Robert Wilson of the Bell Telephone Laboratories. The measured temperature turned out to be about 2.7 kelvins, proving that the early theoretical work had been essentially correct. Neither Penzias nor Wilson, nor any of the other participants in the first measurements, had ever read the papers of Gamow and his coworkers. Theirs had been a discovery without a context.

The history of the subject of this chapter, the unification of electromagnetism and the weak interactions, includes, as we shall see, several remarkable examples of premature theoretical discoveries. We recall that when Fermi constructed it, his theory of β-decay was built on an analogy with electromagnetism according to which the interactions take place between currents of charged particles. Yukawa made the analogy still closer when he introduced the pi-meson as the carrier of the weak force between the currents, just as the photon carries the electromagnetic force between the currents of charged particles. However, the pi-meson, as the carrier of the weak interactions, fell out of favor in the late 1940s when it was realized that the weak interactions had a universal character.

"Universal character" means that both the hadrons, as manifested in a β-decay like $n \rightarrow p + e^- + \bar{\nu}_e$, and the leptons, as manifested in a decay like $\mu^- \rightarrow e^- + \bar{\nu}_e + \nu_\mu$, exhibited weak interactions of about the same order of magnitude. Thus the carrier could not be the pi-meson, which has both strong and weak interactions; instead it must be some sort of particle that has only weak and electromagnetic interactions. Apart from this, it turned out that the pi-meson, which is spinless, could not provide a form of the weak interaction that fitted the newly growing body of experimental data. The form of quantities, such as the β-spectrum, is sensitive to the spin nature of the carrier. After some experimental confusion, it became clear by the late 1950s that the spin of the carrier had to be 1, just like that of the photon. It couldn't literally be the photon, since the photon is massless and gives rise to the long-range Coulomb force and the weak interactions are extremely short range. Thus arose the notion of the weakly interacting, massive vector, or spin-1, meson as the hypothetical carrier of the weak interactions.

The first suggestion that this weak vector meson might be a gauge particle—a photon-like object introduced to preserve some form of local gauge invariance—appeared in 1957 in a series of papers and lectures by Julian Schwinger. Schwinger's papers were yet another example of discovery without context, and it was several years before they were appreciated. To understand some of the features of Schwinger's picture, we must introduce a distinction that will play an important role in what follows: the distinction between charged and neutral currents.

In a purely electromagnetic process, like electron scattering, there is no change of electric charge when the charged particles emit and absorb photons (Figure 6-1). But in a process like β-decay, as Figure 6-2 shows, each time the particles emit or absorb a weak meson there is a gain or loss of electric charge. One could, Schwinger emphasized, also imagine weak processes in which there is no exchange of charge such as the one shown in Figure 6-3, which would be carried by a weak electrically neutral vector meson.

Figure 6-1. The emission of a photon by an electron.

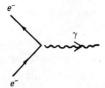

Figure 6-2. The emission of a charged weak meson by a neutron.

Figure 6-3. The emission of a neutral weak meson by a neutron.

At the time that Schwinger wrote his paper, there was no evidence for neutral weak interactions of this sort. On the contrary, it seemed as if they were ruled out, since processes like $K_s^0 \to \mu^+ + \mu^-$ had never been observed. Hence, in Schwinger's original model, the photon and the positive and negatively charged weak vector mesons, the W^+ and W^-, were the only gauge mesons introduced. The gauge symmetry was taken to be the isotopic spin group, SU(2). Even though Yang and Mills had published their paper 2 years earlier, Schwinger did not use local SU(2) invariance. Using it would have led him to a Yang-Mills version of his theory, with its massless gauge particles. What he did was to put in, more or less by hand, the three gauge mesons in a triplet representation of SU(2) with, in the limit of the exact symmetry, a common zero mass. But the masses of the W^+ and W^- could not be zero and still mediate the observed weak interactions.

What to do? Here Schwinger introduced the first intimations of an idea for generating masses that, although it was certainly not recognized at the time, would reverberate in elementary-particle theory for the next three decades. Here is how Schwinger gave his W^+ and W^- their mass. In addition to the gauge fields, he introduced a new field corresponding to a spin-0 particle called by him the σ-meson. When this particle is coupled in a suitable fashion to the W's, part of the field looks just like the effect of a mass, and that is what Schwinger assumed it was. In technical language, Schwinger assumed that the σ-field had a nonvanishing constant average value in the vacuum, and that constant value he related to the mass. It was by no means the theory we now use, since Schwinger did not discuss how the σ-field might acquire such an average vacuum value, but it was an intimation.

Among the people on whom Schwinger's papers had an impact was Sheldon Glashow. That is not too surprising, since Glashow was Schwinger's doctoral student and, indeed, wrote his thesis on using gauge theories to unify electromagnetism and the weak interactions. Glashow was concerned about the problem of renormalizability. The electromagnetic interaction was and the weak interaction—the Fermi theory—wasn't. It did not make any sense, therefore, to put them in a union unless that improved the renormalizability of the combined theory. But it became clear that if one simply grafted the masses on by hand, the unified theory with massive weak vector mesons was not renormalizable. Nonetheless, in 1961, Glashow introduced a new version of a unified weak and electromagnetic theory based on an enlargement of the SU(2) group. (The same theory was rediscovered independently a few years later by Abdus Salam and John Ward.)

In the new theory there is the combined invariance of two groups, SU(2) and SU(1). The latter is the simplest unitary group consisting of just one-dimensional changes of phase. That is the gauge group that generates ordinary electromagnetism. The ingenuity of Glashow's choice of groups was that it enabled him to include parity nonconservation for the weak interaction sector of his theory. The combined group is written as SU(2) × U(1). This group evokes four vector gauge mesons. Two of them can be identified with the W^+ and W^-, the third with the photon, and the fourth, which is also electrically neutral like the photon and which is called the Z^0, is something new. The Z^0 interacts weakly in a parity nonconserving way, whereas the photon interactions conserve parity. Once again there was the problem of the various masses. On this, one finds in Glashow's paper the following remarkable statement:

> The mass of the charged intermediaries must be greater than the K-meson mass, but the photon mass is zero — surely this is the principal stumbling block in any pursuit of the analogy between hypothetical vector mesons and photons. It is a stumbling block we must overlook.

The masses having, once again, been put in by hand, the theory is, once again, nonrenormalizable. To add to the general ambiance, the presence of the Z^0 leads to just the sort of weak neutral currents which had not been observed and which Schwinger's model was constructed to avoid. Apart from a certain mathematical elegance, the theory did not at the time seem to have much else to recommend it. It nonetheless, as the rest of this chapter will explain, turned out to have been right.

At this point the scene shifts, at least for a few years, from the weak interactions to the general concept of broken symmetry. The two will come back together in 1967. Until the 1960s, the canonical example of a broken symmetry, and how it was to be interpreted, was given by the isotopic spin, the group SU(2). The presence of such a symmetry was suspected because the observed particles seemed to fall into multiplets whose masses differed by small amounts relative to the mass of any of the members. One suspected that electromagnetism was the culprit in breaking the SU(2) symmetry, since the mass differences within the multiplet were given, in order of magnitude, by e^2m, where m is any one of the multiplet masses. To express this symmetry and its breaking, we begin with a set of equations in which the nonsymmetric part of the interaction is dropped. The resulting solutions then exhibit the exact multiplet structure. Now, we add a small nonsymmetric perturbing interaction to the original equations. The solutions then will not be multiplets degenerate in mass; instead, they should be sufficiently close to the original multiplets that one recognizes, by looking at it, that the solution has come about from a small perturbation in the neighborhood of the symmetric solution. This realization of symmetry, which is as old as quantum mechanics itself, often goes under the name of the "Wigner-Weyl realization" because Eugene Wigner and Hermann Weyl played such a significant role in its mathematics.

In the early 1960s, however, Yoichiro Nambu and his coworkers introduced a new concept of broken symmetry into elementary-particle physics. It became known — for reasons that are not entirely clear — by the somewhat unfortunate name of "spontaneously broken symmetry." In this kind of symmetry, the equations retain their symmetric character in the presence of *all* the interactions. In fact, the symmetry is realized not from the solutions, but only through the symmetric character of the equations, which need not exhibit anything that looks like the traces of a multiplet structure. A key question here is the symmetry of the solution of lowest energy, the so-called vacuum state. In the Wigner-Weyl realization the vacuum state is a state with a definite symmetry. In the case of the isotopic spin, for example, it is a state with zero isotopic spin, a singlet. On the other hand, in a theory with spontaneously broken symmetry the vacuum state is not a state with a definite symmetry. Indeed, it would be more descriptive to call this type of symmetry breaking vacuum-broken symmetry.

A few interesting applications of that idea were made by Nambu and his collaborators in strong interaction physics, but then the roof fell in. In 1960, responding to Nambu's idea, Jeffrey Goldstone published a paper which seemed to kill it, at least as a viable symmetry-breaking method for elementary-particle physics. In the context of a simple explicit model which exhibited spontaneous symmetry breakdown against SU(2), Goldstone showed that that theory must necessarily contain a *massless* spinless particle that was the dreaded Goldstone boson, "dreaded" because no such massless particle had ever been observed. In 1979, on the occasion of the presentation of the Nobel prizes in physics, Steven Weinberg, who shared the prize that year with Sheldon Glashow and Abdus Salam, recalled his feelings when he first learned about spontaneous symmetry breaking only to learn, subsequently, about the Goldstone boson. He writes:

> As theorists sometimes do, I fell in love with this idea [spontaneous symmetry breaking]. But as often happens with love affairs, at first I was rather confused about its implications. I thought (as it turned out, wrongly) that the approximate symmetries — parity, isospin, strangeness, the eightfold way — might really be exact *a priori* symmetry principles, and that the observed violations of these symmetries might somehow be brought about by spontaneous symmetry breaking. It was therefore rather disturbing for me to hear of a result of Goldstone, that in at least one simple case the spontaneous breakdown of a continuous symmetry like isospin would necessarily entail the existence of a massless spin zero particle — what would today be called a Goldstone boson. It seemed obvious that there could not exist any new type of massless particle of this sort which would not already have been discovered.

Weinberg goes on:

> I had long discussions of this problem with Goldstone at Madison [Wisconsin] in the summer of 1961, and then with Salam while I was his guest at Imperial College in 1961–62. The three of us soon were able to show that Goldstone bosons must in fact occur whenever a

symmetry like isospin or strangeness is spontaneously broken, and that their masses then remain zero to all orders of perturbation theory. I remember being so discouraged by these zero masses that when we wrote our joint paper on the subject, I added an epigraph to the paper to underscore the futility of supposing that anything could be explained in terms of a noninvariant vacuum state: it was Lear's retort to Cordelia, "Nothing will come of nothing: speak again." Of course, *The Physical Review* protected the purity of the physics literature, and removed the quote. Considering the future of the noninvariant vacuum in theoretical physics, it was just as well.

There matters stood until 1964. In that year, a group of papers that appeared showed that there might be a way out—a rather *outré* way out but, nonetheless, a way out. One of the first of these papers was a brief note written by the British theorist Peter Higgs. In his note, Higgs showed that there were theories of one class in which the general proof for the existence of Goldstone bosons, as an adjunct to spontaneous breaking, itself broke down. These were theories with gauge fields coupled to conserved currents such as electrodynamics or the Yang-Mills theories. (They had first been noted by the condensed matter theorist Philip Anderson.) The reason for this breakdown is very subtle and not readily put into nontechnical language. It comes down to the fact there is an apparent clash between the requirements of gauge invariance and those of relativity in such theories. At a deeper level, this apparent contradiction disappears, but along the way it gives rise to additional mathematical terms that spoil—fortunately—the connection between spontaneous symmetry breaking and the Goldstone bosons.

Even better news was to follow. This was the work of several theorists including Higgs, who alluded to it at the end of his original note. To understand this wonderful new result, we may consider the original example on which it was worked out: the electrodynamics of a massless charged scalar meson. This theory exhibits a local gauge symmetry, $U(1)$. We may break this symmetry "spontaneously" by demanding that this scalar field have a nonvanishing constant average value in the vacuum. It is this demand that, in the other theories, conjured up the Goldstone boson. Here too, a Goldstone boson is conjured up, but something astonishing happens: The boson is not coupled to anything. A massless particle that isn't coupled to anything is irrelevant, and it can just be factored out of the theory. But something still more astonishing has happened. The "photon" that was massless has now acquired a mass. Put in another way, the "electrodynamics" of a charged massless boson, with a spontaneously broken gauge symmetry, is not really electrodynamics at all. It is, rather, a theory of an uncoupled, irrelevant, massless boson and a massive vector meson. This totally unexpected result turned out to be the key to unifying the weak and electromagnetic interactions. That was to happen in 1967.

In his Nobel address, Weinberg reports that, in the fall of 1967, "I think while driving to my office at M.I.T., it occurred to me that I had been applying the right ideas to the wrong problem." Weinberg had been using the Higgs mechanism to try to give masses to the strongly interacting vector mesons such as the ρ. Now he realized he had at hand the perfect tool to unite the massless photon and the massive weak vector mesons in some spontaneously symmetry-broken manner. What he did was to use the group SU(2) × U(1), first introduced by Glashow, along with the Higgs idea of a spontaneously broken gauge symmetry. Here is how it worked. The electron, because it has a mass, has both a left-handed and right-handed part, whereas the neutrino has only a left-handed part. Thus one can unite the left-handed part of the electron and the neutrino into a doublet $\begin{pmatrix} e_L \\ \nu_L \end{pmatrix}$ and the right-handed part of the electron, e_R, is a singlet. The neutrino, being massless in this model, has no right-handed part to go along with its left hand. One then demands that the theory be invariant against *local* transformations of the combined group SU(2) and U(1). This, as Yang and Mills had shown, requires the existence of the massless vector gauge quanta. There are four of them: three for the SU(2) group and one for the U(1) group.

That much had been in the Glashow, Salam, and Ward papers. The novelty came in how the masses were introduced. To that end, Weinberg introduced a doublet of scalar fields—Higgs mesons—$\begin{pmatrix} \phi^0 \\ \phi^+ \end{pmatrix}$. The field ϕ^0 is allowed to have a nonvanishing constant average value in the vacuum. That triggers the spontaneous symmetry-breaking mechanism. The two charged weak mesons, the W^+ and the W^-, acquire a mass. The two original neutral vector gauge mesons become two new neutral vector mesons, one of which acquires a mass and the other of which does not. The one that acquires a mass is identified with the weakly interacting Z^0, and the massless vector meson is identified with the photon. The ϕ^0 particle also acquires a mass and should be observable as a weakly interacting meson. There are two dimensionless coupling constants in the theory, g and g'—two, because the gauge group of the theory is the product group SU(2) × U(1). The ratio of these constants is a pure number—a free parameter—in the theory. It is customary to define it as the tangent of an angle, the so-called Weinberg angle: $g'/g = \tan \theta_W$.

To understand how the dimensionless constants g and g' are to be found from experiment, we need to return to the distinction between the local and the nonlocal theories of the weak interactions as discussed in Chapter 1. The point is best made in terms of Feynman diagrams. Let us consider the process $\nu_e + e \rightarrow \nu_e + e$—elastic electron-neutrino

Figure 6-4. Elastic electron-neutron scattering as described in the local Fermi theory with the Fermi coupling constant G_F.

scattering. In the original Fermi theory this is described by Figure 6-4. Let us contrast this with the diagram for the same process as mediated by a charged W. There is also a diagram (Figure 6-5) involving the Z^0,

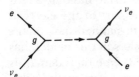

Figure 6-5. Elastic electron-neutrino scattering as described in the intermediate vector meson theory with the dimensionless constant g.

which we will ignore for this purpose. If the energies of the electron and neutrino are not large compared to the mass of the W, we expect the two descriptions to give sensibly the same answer. That enables us to relate the constant g to the dimensional constant G_F. Because the latter has rather peculiar dimensions, the relation looks a little odd. It is essentially

$$\frac{G_F c}{(h/2\pi)^3} \simeq \frac{g^2}{m_w^2}$$

Thus if we knew g, we would know the mass of the W, and vice versa. We can, however, raise the following point. If electromagnetism and the weak interactions are to be truly unified by this theory, we would expect that the coupling constants g and e would be of the same order of magnitude. We can use that condition by demanding, hypothetically, that g and e be precisely identical and asking what sort of mass for the W that would imply. It turns out that it would be about 40 GeV. Hence, if this theory has anything to do with reality, we would expect a W, depending on how g and e are related experimentally, with a mass of at least 40 GeV mass and a Z^0, which the theory predicts to be substantially heavier.

Weinberg published all this in 1967, and for the next 4 years it was almost totally ignored. Weinberg later noted that, during that time, his paper was cited only five times, including his own references to it. Why? Probably there were several reasons. Although the initial unbroken theory looked rather simple, the broken theory, which was to be compared to experiment, did not. It had two arbitrary constants, to say nothing of an observable massive Higgs meson that no one had observed. It also had neutral currents that no one had observed either. Furthermore, it was not even clear that it was renormalizable. To that end, all Weinberg could offer, in his 1967 paper, was the prospect that it *might* be, since the unbroken theory was generally assumed to be and since the masses were introduced in this very special way.

There things stood until 1971 when, enter Gerardus t'Hooft. In 1971, t'Hooft was a graduate student in Utrecht in Holland. His professor, Martinus Veltman, had been working on the problem of trying to renormalize the massive Yang-Mills theories for several years, and he communicated his concerns to t'Hooft. In fairly short order, t'Hooft showed that the massless theory was renormalizable, a result that was widely believed but for which he supplied a formal proof. He then proceeded to rediscover the Higgs mechanism for himself. With the Higgs mechanism in hand, he turned to the renormalization of gauge theories with spontaneous symmetry breaking. He took as his example a general Yang-Mills theory with the gauge bosons coupled only to themselves, reasoning that, if that theory could be renormalized, adding leptons would not change matters dramatically.

The point about the spontaneous symmetry breaking is that enough of the original gauge symmetry is maintained that one can make gauge transformations that get rid of the most divergent expressions. That is just what could not be done if the masses were put in by hand and the gauge invariance destroyed that way. t'Hooft then went on to show that the Weinberg model, with the leptons, also was renormalizable. All that was published in 1971 in a 27-page paper which almost no one understood or thought was right. The methods were quite novel, and t'Hooft was a completely unknown graduate student who was claiming to have solved a very deep field theory problem that had defeated everyone else for nearly two decades.

t'Hooft's paper, which turned out to be correct, was so striking that it set theorists off in two distinct directions. One group, notably the late Benjamin Lee and his collaborators, was able to prove the renormalizability of the theory by employing methods that were more familiar to theorists. On the other hand, people began computing physical processes with the aid of theory. That was equally important because it supported confidence that all the complicated terms in the theory were both necessary and sufficient to render it sensible. To give a typical example, consider the difficult-to-observe but physically possible process $\nu + \bar{\nu} \to W^+ + W^-$. If the rate for this process were computed by using W-meson theory in which the masses had been put in by hand, the answer found would have violated the conservation of probability. The rate increased without limit as the energy of the incident neutrinos rose. This was a very clear indication that the theory was sick. On the other hand, when the unified SU(2) × U(1) theory with all its strange-looking terms involving Higgs mesons and the like was used, the offending terms exactly canceled. It all seemed too good to be true.

For awhile it appeared that it *was* too good to be true. This was the matter of the neutral currents. In the Weinberg version of the unified theory it was absolutely essential to have neutral currents. The currents coupled directly to the Z^0 and led to diagrams for electron-neutrino scattering such as the one shown in Figure 6-6. There were versions of

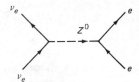

Figure 6-6. Neutrino-electron scattering with the exchange of a Z^0.

the unified theory which did not have neutral currents, but they seemed a little less natural. To explore the consequences of the neutral currents for the full range of the weak processes, including those that involved hadrons, it was necessary to extend the theory to include hadrons as well. That was done in 1971 by Weinberg in the context of the quark model. To appreciate this extension, we must backtrack a little. By the late 1950s it was clear from experiments that weak processes in which the strangeness was changed, such as $\Sigma^- \to \Lambda^0 + e^- + \bar{\nu}_e$, were systematically weaker than processes, such as $n \to p + e^- + \bar{\nu}_e$, in which the strangeness did not change. That had been incorporated into the theory in a very elegant way by the Italian theorist Nicola Cabbibo in 1963. He introduced what is known as the Cabbibo angle, θ_c. In terms of this angle we can write, symbolically, the three currents—ordinary, strange, and leptonic—that take part in the weak interaction in the form

$$J = \cos \theta_c \, O + \sin \theta_c \, S + L$$

From experiment, $\theta_c \approx 0.26$ radian, a mysterious number whose origins are still not clear. From the point of view of the quark model, these currents are made out of quarks. A β-decay process, for example, is pictured as a process in which one quark turns into another—a change of flavor—with the emission of one of the weak vector mesons. If these currents are to be incorporated into the SU(2) × U(1) model, the quarks must be so arranged that they are representations of the SU(2) × U(1) group as well. As Weinberg showed, that could be done, beginning with the three original quarks, as follows: First we write the lepton doublet with its electron-right-handed singlet part:

$$\begin{pmatrix} \nu_{eL} \\ e_L \end{pmatrix} \quad \text{and} \quad e_R$$

Next we invoke a very powerful, but still rather mysterious, principle which has come to be known as the quark-lepton analogy. Quarks and leptons seem to imitate each other as far as the *weak* interactions are concerned. (Color, which is a strong interaction concept, has no leptonic analog of which we are aware.) To mimic the leptons and to take into account Cabbibo's angle, we write the charmless quark sector in the form

$$\begin{pmatrix} u_L \\ d_{cL} \end{pmatrix} ; u_R, d_{cR}$$

where

$$d_c = \cos \theta_c \, d + \sin \theta_c s$$

is the Cabbibo rotated quark. We can put the content of this rotation in a slightly different, although equally mysterious, way. The quarks with a definite mass, in this case the u, d, and s quarks, are not the same as the quarks that couple to the weak mesons; they differ by a rotation in quark space through the Cabbibo angle θ_c. Why that should be so is something we do not really understand in a deep sense as yet. That is the way the electron sector is constructed.

What about the muons? First, we begin with the leptonic piece which, unsurprisingly, is written

$$\begin{pmatrix} \nu_{\mu L} \\ \mu_L \end{pmatrix} ; \mu_e$$

We now see that the existence of the charmed quark has opened up the beautiful possibility of writing its lepton analog as

$$\begin{pmatrix} c_L \\ s_{cL} \end{pmatrix} ; c_R, s_{cR}$$

where s_c is the Cabbibo rotated strange quark defined as

$$s_c = -d \sin \theta_C + s \cos \theta_C$$

The reader will have noted the change in signs; it incorporates the GIM mechanism discussed in the preceding chapter. For example, let us consider the leading contribution to the process $K_L \to \mu^+ + \mu^-$. It consists of the two graphs shown in Figure 6-7. If the masses of the u and the c quark are taken to be the same, these two graphs in Figure 6-7 exactly cancel. In general, this mass difference can be taken small enough to make the process negligible, which is the GIM mechanism. There are in this theory no neutral strangeness-changing currents.

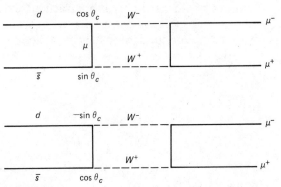

Figure 6-7. The leading quark-model contributions to the decay of the K_L^0 into a $\mu^+ - \mu^-$ pair. In this model the K^0 is bound $d\bar{s}$ pair.

In 1975, the quark-lepton analogy took a giant step forward, or so it became clear in retrospect. That year a group at SLAC headed by Martin Perl found e^+e^- collisions that produced correlated pairs of electrons and muons. That was to be expected if the e^+e^- collision had produced a new heavy lepton pair which was decaying into the familiar leptons. Indeed, that is what it turned out to be. The new, unexpected lepton, was named the τ. It has a mass of 1784 MeV and a lifetime of only about 3×10^{-13} second. In its decay into the light leptons, say, $\tau \to \mu^- + \bar{\nu}_\mu + \nu_\tau$, a new neutrino, which is called ν_τ for obvious reasons, is emitted. It has been shown at least indirectly by experiment that this neutrino is distinct from both the ν_μ and the ν_e.

Believers in the quark-lepton analogy were not shocked when, as we mentioned in the preceding chapter, evidence for a new quark, the b-quark, turned up in 1977. If it is going to fit into a doublet structure, like its predecessors, it needs a partner, which is why people are expecting the t, or top, quark to appear as soon as the machines become energetic enough. As was first pointed out by Minoru Kobayashi and Toshihide Maskawa in 1972, before any of the experimental discoveries, a new possibility opens up with these flavors. We can imagine a more general rotation in the quark space than the one envisioned by Cabbibo but one which includes his as a special case. This rotation involves all three quarks, d, s, and b. In this picture the three SU(2) doublets can be written as

$$\left(\begin{matrix} u_L \\ \alpha \, d_L + \beta \, s_L + \gamma \, b_L \end{matrix}\right), \left(\begin{matrix} c_L \\ \alpha \,'d_L + \beta's_L + \gamma'b_L \end{matrix}\right), \left(\begin{matrix} t_L \\ \alpha \,''d_L + \beta''s_L + \gamma''b_L \end{matrix}\right)$$

where the α, β, and γ characterize this more general rotation. The beauty of this generalization is that it can accommodate CP, or time reversal, noninvariance in a very elegant way. Not all the α, β, and γ need be real numbers. If the quarks are suitably defined, it can be shown that one complex phase is left over, along with the three real angles of rotation. This complex phase embodies the CP violation. Its origin still needs to be explained, but this way of looking at things makes precise just what does need to be explained.

Now that we have discussed the theory for the weak interactions of the hadrons as well as the leptons, we are in a position to discuss the matter of the experimental consequences of the neutral currents. There are two classes of processes to be considered: the processes which cannot take place at all, except in higher orders in the weak interactions, if the neutral currents are absent and the processes that take place electromagnetically, to which the weak neutral currents make a small but measurable contribution. We shall give examples of both kinds, beginning with the first. Imagine we have a high-energy beam of muon neutrinos impinging, say, on a bubble chamber. What sort of things can happen? First there are the reactions, $\nu_\mu + e \rightarrow \mu^- + \nu_e$ and $\nu_\mu + n \rightarrow p + \mu^-$, which take place with the exchange of the charged W as shown in Figure 6-8. These processes take place in a theory with no neutral currents. But in a theory with neutral currents there can be, to

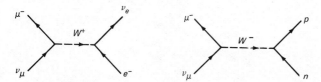

Figure 6-8. Charged current contributions to ν_μ inelastic scattering.

the same order in the weak interactions, muonless reactions such as $\nu_\mu + p \rightarrow \nu_\mu + p$. In a bubble chamber picture, these events will appear as a thick track with no connection to any charged particle. They can, unfortunately, become confused with collisions that are initiated by neutrons, and that makes the experiment more difficult. If there are no neutral currents, such neutrinoless events take place through only higher-order diagrams such as Figure 6-9. In addition, there can be

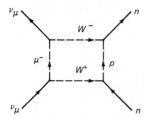

Figure 6-9. A higher-order diagram that mimics a neutral current.

neutral current events of the form $\nu_\mu + p \rightarrow \nu_\mu + X$, where X represents some strongly interacting set of hadrons. In 1973 a CERN group, by using the huge Gargamelle bubble chamber and after analyzing some 300 000 events, found a couple of hundred events that could be attributed to the neutral current. It was a triumph for the theory. The weighting of the charged and neutral components of the weak current depends upon the magnitude of the Weinberg angle. The best present fit to all the data is an angle given by $\sin^2 \theta_W = 0.226 \pm 0.004$. Why the angle should have just that magnitude is still a mystery.

The other class of neutral-current phenomena involves processes that would happen anyway through electromagnetism but are slightly "contaminated" by the presence of the weak interaction through the neutral current. The diagrams in Figure 6-10 for electron-electron scattering il-

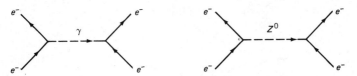

Figure 6-10. Electromagnetic and weak diagrams contributing to electron scattering.

lustrate the general point: The contamination shows up as a violation of parity which can take place only because of the weak interaction contribution generated by the Z^0. Experiments to find this are very difficult because the effect shows up, compared to the dominant parity-conserving electromagnetic scattering, at a level of about 1 part in 10 000. Nonetheless, it has been observed. The most celebrated experiment was done in 1978 at SLAC. There a group used polarized beams of electrons to scatter from deuterium. By "polarized" one means that the electron spins are made to point predominantly either in the direction of their momenta or oppositely. If parity were conserved as would be expected in a purely electromagnetic process, scattering would be the same for either sense of polarization. But by reversing the electron polarization, the group discovered a difference between the left- and right-handed electrons in the scattering process at just the level that the theory had predicted; another triumph.

These various successes would have been hollow indeed if no one had been able to find the physical W's and Z's. Unlike the quarks, these must be observable particles. Their discovery at CERN in 1982–1983 is a sort of apotheosis of modern high energy experimental physics. Nothing that was done, from the fantastic accelerator design to the computers, would have been remotely conceivable to, say, Rutherford. The first thing to appreciate is the accelerator. To understand what was done, we first note what needed to be done. The unified theory predicts that the mass of W obeys the simple equation

$$m_W \simeq \frac{37\ GeV}{\sin \theta_W}$$

which, with the value 0.475 for $\sin \theta_W$, gives a W mass of about 80 GeV. Let us suppose that we want to produce W's in the reaction $p + \bar{p} \to W^+ + X$, where X is some collection of particles of mass small compared to that of the W. If we use a fixed laboratory target, then the kinematics shows us that we would need a machine with an energy of some 3000 GeV, that is, 3 TeV. No such machine exists, and certainly none existed when the plans for this experiment were being made. The situation for the Z^0 is even worse, since by use of the same Weinberg angle it is predicted to have a mass of some 90 GeV. In 1976, Carlo Rubbia and Peter McIntyre from Harvard and David Cline of the University of Wisconsin, all of whom were working at CERN, suggested a very daring program: to convert the various CERN accelerators either existent or under construction into entities that would function as proton-antiproton colliding-beam accelerators. That was daring because, if accepted, it meant committing the entire experimental effort of a laboratory to a project that might fail.

At the time, CERN had two existing machines: the intersecting storage rings (ISR) and the proton synchrotron (PS). The latter is a 28-GeV synchrotron which was then used to inject accelerated protons into the ISR, where individual beams circulated in opposite directions to provide a proton-proton colliding-beam machine. Under construction then was the super proton synchrotron (SPS), which would have the capacity of accelerating protons to 270 GeV. In addition, there was a brilliant technological development at CERN that involved taking a collection of protons or antiprotons, injecting them into an accelerator with a wide spread of momenta, and "cooling" them so that the energy spread was progressively reduced. The method is due to the Dutch physicist Simon van der Meer, who called it stochastic cooling. Stochastic, or random, cooling involves sampling slices of the beam randomly and making a large number of small corrections which nudge the beam into a uniformity of energy. By 1976 the method had been successfully tested at the ISR, where it was shown to reduce the spread of the beam, both in space and momentum, by something like 10 percent.

Here, then, is the final extraordinary scenario. Step 1 consists of letting a 26-GeV proton beam from the PS strike a metal target and produce antiprotons with energies centering around 3.5 GeV. Each pulse of the PS contains about 10^{13} protons, which yields about 10^7 antiprotons. The antiprotons are guided into a machine called an antiproton accumulator. It had to be newly built, and it came into operation in 1980. The machine is roughly square-shaped, and in it the antiprotons circulate and "cool." It is constantly being fed pulses of antiprotons from the PS. It takes *40 hours* of this stacking before some 10^{12} antiprotons are accumulated. The antiprotons are then fed back into the PS, where they are accelerated to 26 GeV. At that energy they are fed into the SPS, where they can be accelerated to 270 GeV in a counter sense to the 270-GeV protons which are circulating in that machine. The whole thing, of which a sketch is shown in Figure 6-11, is a miracle of engineering.

In the sketch we have shown the two underground areas, UA1 and UA2, where detectors are located. The detectors are enormous. The UA1 detector complex, for example, weighs 2000 tons and was built in an underground garage next to the beam, from which it is rolled on tracks to its active position athwart the beam. The central detector in the complex is a cylinder 19 feet long and 7.5 feet in diameter. The proton-antiproton colliding beams run through the center of the cylinder in an evacuated pipe. Unlike the bubble chamber, this detector does not produce direct visual images that can be photographed; instead, it produces a series of electric pulses that are fed into computers that reconstruct the events. To give some notion of the complexity of this operation, the UA1 detector alone employs nearly 150 physicists and engineers. An experimental paper from the group probably contains

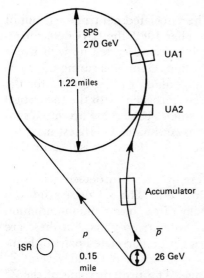

Figure 6-11. The p and $\bar{p}$ beams entering the SPS after having been accelerated in the PS. The antiprotons are stored first in the accumulator. UA1 (underground area 1) and UA2 are the detector sites.

more names than the total number of people who ever worked with Rutherford at Cambridge and Manchester combined.

How, then, is this machinery used to detect W and Z^0 particles? If we think of the $p\bar{p}$ collision in terms of the quark model, the most likely W- and Z^0-producing collisions are

$$u + \bar{d} \rightarrow W^+ + X$$

$$d + \bar{u} \rightarrow W^- + X$$

$$u + \bar{u} \rightarrow Z^0 + X$$

$$d + \bar{d} \rightarrow Z^0 + X$$

where, in all cases, X is a hadron jet. Compounding the detection problem is the fact that the overwhelming number of $p\bar{p}$ collisions are not W- or Z^0-producing but simply hadron-producing. Finding the rare W- and Z^0-producing collisions is like finding a needle in a hayfield — a haystack would be relatively simple.

What are needed are events with very characteristic "signatures." In the case of the charged W's, which can decay into either muons or electrons, that is, $W \rightarrow \mu$ (or e) $+ \nu$, the signature is a hadron jet correlated to a very energetic charged lepton. The associated neutrino gets out of the detector. In a 30-day period, November 1982, the UA1 detector recorded a billion proton-antiproton collisions and ended up with five — five! — candidates suitable for W production. Upon measuring the energies, the experimenters came up with a value of 81 ± 5 GeV for the mass of the putative W, a figure entirely consistent with the theoretical prediction and the observed Weinberg angle. UA2 came up with a similar number of events and a similar mass estimate. The latest mass for the W is measured to be 81.8 ± 1.5 GeV, and the width is measured to be less than 6.5 GeV.

In the spring of 1983, the groups were back at their detectors looking for the Z^0. Here the signature is different. The Z^0 decays into an electron-positron pair and, about equally often, into a muon-antimuon pair. It also has decays into neutrino-antineutrino pairs, but these are not detectable. Hence, the detectors trigger on correlated pairs of high-energy charged leptons. By July, both groups had found several Z^0 events with masses in the 92-GeV range. The present value of the Z^0 mass is 92.6 ± 1.7 GeV, and the width is less than 4.6 GeV. This value of the Z^0 mass is consistent with the prediction of the unified theory using the same Weinberg angle that fits the rest of the data. In 1984 Carlo Rubbia and Simon Van der Meer shared the Nobel prize in physics, and in a deeply symbolic way, this signaled the successful end of a certain era in the physics of elementary particles.

7
The Old
and the New

In addition to his extraordinary imaginative scientific work, the late George Gamow — he died in 1968 after the Big Bang radiation had been discovered — also wrote wonderful popular science books. A generation of scientists got their first taste for science by reading Gamow's popular books. Among those books were the Mr. Tompkins series, some of them illustrated by Gamow himself. Of that artistic enterprise he wrote, "The pictures that have resulted from this bold effort may look somewhat crazy, but after all the stories themselves are rather crazy too." The best description of Mr. Tompkins, and what befalls him, comes from his creator. Gamow writes:

> He is just an ordinary person, a bank employee to be exact, who ventured to attend several semi-popular lectures on such basic problems of modern physics as relativity and the quantum theory. The startling statements he heard at the lectures, one apparently more paradoxical than the next, made such a deep impression on our hero's mind that his sleep was often disturbed by the craziest of nightmares related, if sometimes rather distantly, to what the professor had said. Mr. Tompkins found himself riding a bicycle on the shrinking streets of a weird relativistic city, and hunting on an elephant in a quantum jungle only to be attacked by the tiger from all sides at once, and in other nerve-racking predicaments.

Gamow goes on, with uncharacteristic reticence:

> It is hard to say whether Mr. Tompkins really got anything out of his unusual adventures beside his happy marriage to the professor's daughter Maud and his present dubious privilege of being exposed to his father-in-law's lectures morning, noon and night.

Be that as it may, one can imagine what adventures Gamow might have invented for the hapless Mr. Tompkins if he had been around to see what the elementary-particle physicists and cosmologists have done to his early universe. Gamow, as much as anyone, laid the foundation for the idea that there *was* an early universe; that an early universe could provide an explanation for some things, such as the relic cosmic radiation, that we observe today. Keep in mind that at the time he was doing his work there was a rival theory, the steady-state theory of Fred Hoyle and others, the principal tenet of which was that there was no early universe at all. (Hoyle, it is said, invented the term "Big Bang" as a derisive description of Gamow's activity.) Nonetheless, by modern standards, Gamow's early universe was a pretty drab place. His ylem, the ur-matter out of which everything was to have been built, consisted only of familiar laboratory particles — neutrons, protons, electrons, protons, and neutrinos. In fact, as we shall see, what he called early was not, by the standards of the contemporary cosmologist, early at all. To get a feeling for all of this we are, in the spirit of Gamow, going to transport the unfortunate Mr. Tompkins, in a dream or nightmare, back to around the time of the Big Bang. We will describe what a modern cosmologist would speculate that Mr. Tompkins might see once he got there.

The first question to discuss is how early is early. We believe that the Big Bang, the primordial explosion, took place some 15 billion years ago. Although some very adventuresome cosmologists are willing to speculate about what happened at the instant of the explosion itself, or even before, we shall take a more sedate point of view and start the clock at the so-called Planck time. To understand the significance of that choice, we must backtrack a little. Max Planck, who invented the idea of the quantum of radiation in 1900, had a lifelong interest in "natural units" which would "retain their meaning for all times and for all cultures, including extraterrestrial and nonhuman ones." He realized that with the introduction of his constant h — Planck's constant, which has the dimension of energy-seconds and is a measure of the size of the quantum — it was possible to introduce combinations of constants that would render dimensionless some of the constants of physics, such as Newton's gravitational constant, that seem to have very peculiar or "unnatural" dimensions. We will focus on Newton's constant G_N. As introduced by Newton, this constant has the bizarre dimensions of volume/mass × seconds^2. Why? We begin with Newton's law that says, generically,

$$\text{Force} = \text{mass} \times \text{acceleration}$$

Newton discovered that gravitation can be represented as an inverse-square law, that is, the strength of the gravitational force falls off as the square of the distance between two gravitating masses. The constant of proportionality has the dimensions mass^2 × G_N, where, as before, G_N is Newton's constant. Thus, in this case, Newton's law reads dimensionally

$$\text{Force} = G_N \times \text{mass}^2/\text{length}^2 = \text{mass} \times \text{length}/\text{second}^2$$

If we solve this for the dimensions of G_N, we find, as advertised,

$$G_N = \frac{\text{length}^3}{\text{mass} \times \text{second}^2}$$

In constructing the dimensionless G_N it is customary to use, instead of Planck's constant h, the quantity $\hbar$ defined as $\hbar = h/2\pi$. In addition, one uses the speed of light c. Both of these constants are "universal" in that they do not refer to any particular object. We do not have a universal mass; but if we call the generic mass m, we see that the combination $G_N m^2/\hbar c$ is dimensionless. That serves the role of the dimensionless coupling constant in gravitational processes involving the mass m. The "Planck mass"—a term coined by the physicist John Wheeler—is defined to be the mass which renders this effective coupling constant equal to unity. The Planck mass is the convenient scale on which to measure gravitational effects. How big is it? Not very. More quantitatively, using,

$$\hbar = 1.054 \times 10^{-27} \text{ erg-second}$$

$$c = 3 \times 10^{10} \text{ centimeters/second}$$

and

$$G_N = 6.672 \times 10^{-11} \frac{\text{centimeter}^3}{\text{gram} \times \text{second}^2}$$

we find that the Planck mass $M_p = 2.18 \times 10^{-5}$ gram, about 10 times more massive than a typical speck of dust. If we multiply that by the square of the speed of light c^2, we find the corresponding Planck mass-energy to be 1.22×10^{19} GeV. That energy is huge on the elementary-particle scale. The most energetic accelerator now operating, the so-called Tevatron at the Fermilab, is capable of accelerating protons to "only" 1 TeV, or 10^3 GeV, and even the superconducting supercollider is being designed for a "mere" 20-TeV proton beam.

Corresponding to the Planck energy is a Planck temperature T_p, which is essentially the temperature that a plasma with the Planck energy would have. To go from energy to temperature, we employ

Boltzmann's constant k, which has the value 8.617×10^{-11} MeV/K by using the formula $E = kT$. The temperature is measured on the Kelvin scale, whose absolute zero corresponds to $-273.15°C$. Thus the Planck temperature is 1.42×10^{32} K, to be contrasted to the present temperature of the universe, 2.7 K. A lot of cooling has taken place in the last 15 billion years! Indeed, we shall start Mr. Tompkins off at the Planck time, the time at which the temperature would be the Planck temperature. We can form a time out of the Planck mass and $\hbar$ and c, namely, $t_p = \hbar/M_p c^2$, so that $t_p = 0.54 \times 10^{-43}$ second. Prior to that time, most physicists are not sure about the physics. The effects of quantum gravity, which we do not yet know how to compute very well, may be decisive. So we will put Mr. Tompkins at 0.54×10^{-43} second after the Big Bang and ask what he will observe, beginning with the electromagnetic quanta.

The bulk of the electromagnetic quanta with which Mr. Tompkins has to deal at the Planck time have energies of the order of the Planck energy, 10^{19} GeV. It will be another few hundreds of thousands of years until the temperature drops to a few electron volts and most of the light becomes visible. As we go along, we will use the number of electromagnetic quanta as a benchmark against which to compare the number of particles of other kinds which are around at the same time. The number of electromagnetic quanta is not a conserved quantity, but it is still useful to compare its value at various epochs to the corresponding value of the other particle numbers. At this epoch, it is reasonable to suppose that the relative abundances of all the particles are about equal and are about the same as the abundance of the electromagnetic quanta. Some of the particles Mr. Tompkins will have to confront at the Planck time will be our old friends; some will be old friends in new guises; and some will, we believe, be new friends altogether.

In the old friends category are the leptons. All of the leptons we have discussed, charged and neutral, should be there, along with any additional higher-mass lepton flavors not yet discovered in our terrestrial laboratories. We can divide the neutral leptons into two categories which we can call light and heavy neutrinos. All the neutrinos discovered in the laboratory so far are light neutrinos. Indeed, they may have mass 0. But theorists have proposed various heavy neutral leptons— "neutrinos"—as possibilities. If these are stable, then for reasons we will discuss later, they must have masses greater than a few GeV. If they are unstable, they can be less massive, depending on what they decay into. In any event, if they exist, they are presumably part of the mix at the Planck time. We have no reason to suppose that the number of leptons and antileptons should differ sensibly at the Planck time. Indeed, Mr. Tompkins would see leptons and antileptons mutually annihilating into each other and into photons and the other gauge mesons. This activity will keep the plasma in "equilibrium," meaning, among other things, that the different particle components will be at the same temperature.

Speaking of gauge mesons, there will be the Z^0 and the W's, but there may also be other gauge mesons which we can call X particles. All these particles will be massless; their masses will be acquired later. These X particles might be the participants in a scenario for distinguishing matter from antimatter which we are going to discuss shortly. Speaking further of mesons, there is the question of the Higgs meson, a loose end or loose cannon, left over from the preceding chapter. If the spontaneous symmetry breaking is achieved through the vacuum properties of a real Higgs meson, we would expect such a meson to be around at the Planck time and, indeed, to show up in laboratory experiments as well. There are theoretical arguments that suggest that the Higgs meson responsible for symmetry breaking in the SU(2) × U(1) theory, if it is real, must have a mass greater than about 8 GeV. The upper limit on its mass, using related reasoning, should be less than about 1000 GeV. There are several theoretical assumptions, which we shall discuss briefly in the next chapter, that go into this limit, and we might be surprised by finding a Higgs meson with a different mass, or by not finding one at all, or by finding several.

The Higgs mesons considered here can be produced in lepton-antilepton collisions and also in collisions of the vector mesons with their antiparticles. They can decay, if they are massive enough, into any of these channels. If they exist, they also should be present at the Planck time. In a certain sense we might be better off if the Higgs mesons don't exist as elementary spin-0 particles. In the long run we seek a theory with the minimum number of arbitrary constants. An elementary Higgs field requires new parameters, masses and coupling constants that have to be explained. We can, in principle, still have a spontaneous symmetry breaking without elementary Higgs particles if, in the spirit of Fermi and Yang, we can find suitable building blocks out of which to construct spin-0-like composite objects with nonvanishing average values in the vacuum. A good deal of work, not entirely successful so far, has gone into trying to make a theory that will do this, and perhaps eventually one will come to pass.

Mr. Tompkins should not feel too disappointed if he does not find elementary Higgs particles at the Planck time, but he should feel devastated if he does not find gravitons. The graviton is to gravitation what the photon is to electrodynamics and what the gluon is to chromodynamics. It is the quantum of gravitation. If we are ever to make gravitation into a quantum field theory, we must have gravitons to transmit the gravitational force. These are spin-2 massless particles that, from the Planck time onward, interact only very weakly with objects that have mass-energy. By the Planck time, the plasma of gravitons interacts so weakly with the other particles that it cools and expands with the expanding universe as if the other particles were not there. We are presumably bathed in gravitons at the present time; they are left over, like the microwave radiation, from the Big Bang. No one has figured out how to detect them. Now to the big surprise. The reader will have noticed that no mention has been made about the hadrons—neutrons, protons, mesons and the like. The reason is that, according to present ideas, there *aren't any* at the Planck time. What Mr. Tompkins will see are naked quarks and gluons—yes, *free* quarks and gluons. It will not be for about 10^{-6} second that the quarks become imprisoned as hadrons.

If present ideas are right, Mr. Tompkins will not see much of anything that is new for approximately the next 10^{-34} second. The universe will peacefully expand from a density of some 10^{97} grams per centimeter cubed to some 10^{78} grams per centimeter cubed. In this regime, the mass density varies as the fourth power of the temperature. By comparison, the density of water is, under standard conditions, 1 gram per centimeter cubed. During this epoch the scale factor of the universe, it turns out, increases as the square root of the time, which means that it has increased by a factor of a few hundred thousand. Because the expansion conserves entropy, it follows, as we noted in the preceding chapter, that the temperature decreases inversely with the scale factor. In this regime, it is falling as the square root of the time. At 10^{-34} second it has fallen to about 10^{27} K. The equivalent energy is about 10^{14} GeV. Once the universe has cooled off to that temperature, assuming present ideas are right, all hell breaks loose. To understand that, we must backtrack.

What has now become known as the standard model is the theory of leptons, protons, quarks, W- and Z-mesons, gluons, and possibly literal Higgs mesons which has, as its underlying symmetry structure, the product of the three groups $SU(3) \times SU(2) \times U(1)$, where the $SU(3)$ group acts only on the colored quarks and gluons. From a fundamental point of view there is, however, something quite unsatisfactory about this group structure, even if the resulting theory seems to fit the facts in the limited range of energies that we have at our disposal. The dissatisfaction flows from the number of free parameters such a theory has. Leaving gravitation aside, the standard theory has three distinct coupling constants, usually labeled g_s, the strong constant, and g and g', which are related to the observed weak and electromagnetic coupling constants. A truly unified theory should have one single coupling constant. How, then, can three apparently distinct coupling constants meld into a single "grand unified" constant?

The idea is that in these theories the "constants" are not really constants at all; instead, they are slowly varying functions of energy. That is what Gell-Mann and Low discovered about electrodynamics when they showed that the effective electromagnetic charge increased slowly with energy, and that is what was discovered in 1973 when it was shown that charges associated with non-Abelian groups like g_s and g decrease slowly with energy. That was the first clue to how unification might come about. At a sufficiently high energy the Abelian constant g' could increase in just such a way that it would meet the decreasing non-Abelian constants g_s and g. In the best of all possible worlds that would require a consistency condition, and this consistency condition would determine the Weinberg angle. But if at the unification energy there is a single constant, it should characterize a single gauge group. In other words, at the unification energy there should be one grand unified gauge group with a single constant.

The real problem, then, is to find the grand unified group. That problem is unsolved, although there have been some valiant tries. Perhaps the most valiant of the tries—a model of what one is after—is the SU(5) group introduced in 1974 by Howard Georgi and Sheldon Glashow. On its face, the SU(5) group seems to be the answer to a maiden's prayer. The first of its desirable characteristics is that it contains SU(3) × SU(2) × U(1) as a subgroup. We do not want to lose the features of the standard model in the process of unification. That is a necessary feature of any successful grand unified group. The second feature that it has, and this is common to most of the grand unified schemes, is the nonconservation of baryon number. Quarks can decay, in these theories, into leptons without baryon number being conserved in the process. As we shall explain shortly, this is a desirable feature, provided it does not happen so rapidly as to conflict with the experiments that have already been done on proton decay. It is desirable because we want to give a cosmological explanation of the striking fact that the universe does not seem to contain any appreciable amounts of antimatter.

As far as we know, the most substantial amounts of antimatter found anywhere in the universe are those stored in the accumulator at CERN. The reason that SU(5), and the other putative grand unified theories, violate baryon number conservation is that quarks and leptons appear in the same representation in those theories. In the limit of exact SU(5) symmetry, those particles are massless and are to be treated as members of the various SU(5) multiplets. The X-mesons in the SU(5) theory act to mediate baryon number nonconserving transitions. If we make use of the condition that the three energy-dependent SU(3) × SU(2) × U(1) coupling constants should meet at a unification energy, we deduce a value of the Weinberg angle $\sin^2 \theta_W \simeq 0.21$, which is consistent with experiment. That is one of the triumphs of the SU(5) model. We also deduce that the unification energy is at about 10^{14} GeV, which, by no accident, is the energy at which we left the hapless Mr. Tompkins. Before returning to his situation, we sketch in Figure 7-1 a curve of the cou-

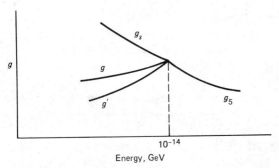

Figure 7-1. The coupling constants as functions of the energy.

pling constants as a function of energy to indicate how the unification can come about in such a theory.

We know from experiment that neither SU(5) nor any other group in which quarks and leptons are unified in that way is the group of the relatively low-energy world we have explored in the laboratory. For perspective, keep in mind that our most powerful accelerator produces protons at energies of 10^3 GeV and even cosmic-ray events are not found with energies in excess of 10^{10} GeV. The clear implication is that the symmetry of the unification group must break down below the unification energy with the result that, after this breakdown, the remaining symmetry is at most SU(3) × SU(2) × U(1). When the temperature cools down so that the equivalent energy is of the order of the mass of the W's or the Z^0, we expect a second symmetry breakdown in which the W's and the Z^0 acquire their masses. At that temperature, the weak and electromagnetic coupling constants are approaching their observed low-energy values.

In the first symmetry breakdown we expect the X particles to take on their masses. Since the only energy scale in the theory is the unification energy, we expect that these masses will be of the order of 10^{14} GeV, well beyond anything accessible to presently contemplated accelerators. We also expect that these symmetry breakdowns will be spontaneous in the sense of the discussion of the preceding chapter. In the grand unified theory there are two, at least two, sets of Higgs particles, some of which acquire a constant vacuum average at about 10^{14} GeV and others at about 100 GeV. That is the mechanism by which the W's, and Z^0, and the like, take on their masses.

At first sight it may seem paradoxical that these vacuum averages can vary with the ambient temperature of the universe. It may seem less peculiar if we recall what we mean by the vacuum state. It is, by definition, the state of the system with the least energy. But that state can be different at different temperatures. It can also be a state with a very complex structure. To appreciate that, we do not have to consider a system as exotic as the expanding universe. Ordinary water will do. When water is above its freezing point, 0°C, its state of lowest energy is liquid water at least until it reaches a 100°C temperature, at which point the state of least energy is steam. Below 0°C the state of least energy is ice. These states have quite different symmetry structures. Steam, being a gas, is the most symmetric of the states, with every molecule being on an equal footing. Ice freezes in crystals each of which has some sort of symmetry which the overall system does not share. Furthermore, when an actual sample of ice is formed, different regions can crystallize with different orientations. Along the boundary of those regions, defects — cracks — can be produced. All those phenomena, including the last, may well have been manifested in the phase changes that the early universe underwent as it passed from regions of higher to regions of lower symmetry.

The matter of the defects has raised both a problem and an opportunity for these models. In the mid-1970s several physicists realized that the cosmological phase changes might be accompanied by the appearance of "defects." That has to do with the fact that when the scalar fields acquire their nonzero vacuum averages, those averages can take on different values in different regions of space. That something like this might take place was suggested by the well-known phenomenon of ferromagnetism. Certain elements, like iron, have the property that, when they are in an environment with a low enough temperature, the atomic magnets in individual spatial regions — domains — can line up with each other. The alignment is generally different from domain to domain, and the boundary regions between the domains can be thought of as defects in the magnetic material. The quantity that determines how these domains are formed is what the solid-state theorists call an order parameter.

In the gauge theories the vacuum averages of the scalar fields serve as the order parameters. The classification of what can happen at the boundary of the different spatial regions in the early universe, as the phase is changing, was given in its generality in a paper by Thomas Kibble in 1976. It brought together the work of a number of previous theorists. There can be, Kibble noted, one-dimensional, two-dimensional, and zero-dimensional defects. The possibility of the latter had been raised by the independent 1974 work of t'Hooft and A. M. Polyakov. They observed that, very generally, when a gauge group breaks down to an SU(3) × SU(2) × U(1) group, by means of Higgs particles of very large mass, stable structures of masses comparable to those of the Higgs mesons that carry units of magnetic charge are inevitably created. These structures are quite different from ordinary magnets which have north and south poles. The new objects have single magnetic poles and are therefore called magnetic monopoles. It has been known, since the speculative work of Dirac in the 1930s, that electrodynamics could not rule out the existence of such objects, even though none had ever been observed. Now the non-Abelian gauge theories seemed to be telling us that they should have been produced in great abundance at the unification energy: a sea of stable magnetic monopoles with masses of the order of 10^{14} GeV. Where are they now? We will come back to this intriguing question shortly.

Some, although fortunately not all, gauge models can produce two-dimensional structures as defects, that is, domain walls. They, it turns out, are more trouble than they are worth. The argument for this judgment is built on a theme to which we will continue to make reference in various contexts. In essence, it is that the walls would be so massive that they would make the universe at large much more "curved" than it is observed to be. Since Einstein's 1916 work on general relativity and gravitation, we have known that gravitating masses make space non-Euclidean—give it a "curvature." Put in less obscure language, if a giant triangle made out of light rays were constructed near a gravitating mass, the sum of the angles of the triangle would not be 360° as in the case of a triangle in Euclidean geometry; it would be a different number. In fact, as the famous solar eclipse experiments which involve the curving of light rays near the sun show, this effect does exist. But it is very small. The universe at large seems quite flat—Euclidean—at least in the present epoch. Hence, theories that predict domain wall structures must be ruled out.

On the other hand, as Kibble observed, the one-dimensional structures—strings with a thickness of some 10^{-3} centimeters but with a weight of 10^{16} tons per inch—might well have existed in the early universe, without doing any damage to present observations. Indeed, as we shall discuss in a moment, they might even do some good. The reason why they are compatible with the observed small curvature is that they didn't stay around very long after they were formed. The theory tells us that primordial strings are either formed as closed loops or must extend across the universe. If the universe is closed—has a finite size—then the strings that expand across the universe also form a loop. In any event, they are under great tension because they have trapped within them regions of very high energy vacuum associated with the grand unified symmetric phase. Because they are under tension, they writhe about, and in so doing they emit gravitational radiation—gravitons—and evaporate. Before doing so, however, they can gravitationally attract clumps of matter to themselves. That is why they might do some good. Cosmologists have been searching for a way to introduce lumpiness into the early universe which, a priori, resembles a featureless gas. The present universe is fairly lumpy, with its stars and galaxies, and that must have come from somewhere, perhaps the cosmic strings. By the way, we must warn Mr. Tompkins to get out of their way if he can, because if one should go through him, the opposite ends of his body will start moving in the direction of the string at several miles a second, hardly a pleasant sensation.

It is also during this very complicated epoch that what is called baryogenesis is presumed to take place. We wish to account for two facts about baryons and antibaryons: that there does not appear to be any significant amount of antimatter and that the number of baryons relative to the number of light quanta is, at present, of the order of 10^{-9} or less. In accounting for these facts, we do not want to invoke any special pleading in terms of, say, initial conditions in the early universe, rigged up just so that they produce these numbers. A plausible explanation should be insensitive to the initial conditions. The general scheme for how to do that was first outlined by Andrei Sakharov in 1967.

With the invention of the grand unified theories, it was possible to produce concrete models of how Sakharov's scenario might work. As we have mentioned, these grand unified theories contain X particles, mesons that mediate baryon-number-violating interactions. When the grand unified theory has its symmetry broken at about 10^{14} GeV, the X particles acquire masses of the order of 10^{14} GeV. The scenario will not work until the X particles become massive. The reason is that, so long as they are massless, no energy is required to reconstitute them after they decay. If we suppose that they decay by a reaction of the form $X \to A + B$, the inverse reaction $A + B \to X$ takes place with equal probability and the decay products never have a chance to accumulate. But the temperature of the universe is continually dropping during that epoch. When it gets below 10^{14} GeV, the decay products do not have enough energy to reconstitute the X's, which can build up. The buildup of X-decay products is necessary but by no means sufficient for baryogenesis. It is crucial that the particles that build up include different numbers of baryons and antibaryons. To see what is involved, let us consider a very simple model in which an X particle has two decay modes, namely, those in which q is a quark and ℓ a lepton:

$$X \rightarrow qq$$

$$X \rightarrow \bar{q}\ell$$

and those in which the antiparticle $\overline{X}$ also has two decay modes:

$$\overline{X} \rightarrow \bar{q}\,\bar{q}$$

$$\overline{X} \rightarrow q\ell$$

These decays fail to conserve baryon number. That is not a surprise, since this is just the kind of decay that the X is supposed to mediate. If we began with an equilibrium situation in which there were equal numbers of X's and $\overline{X}$'s and the rates for the decays of the X's and $\overline{X}$'s were the same, we would get nowhere in establishing a baryon number asymmetry. To make the rates for the X and $\overline{X}$ decays distinct, we must assume that the decays violate both C and CP. These violations, so far observed only in the K^0-$\overline{K}^0$ system, are from a fundamental point of view very mysterious, and it may well be that the key to why they occur is that they are needed to explain why there is no antimatter.

We have now exhibited both the necessary and sufficient conditions required to produce an imbalance between matter and antimatter, although it is a difficult and not completely solved problem to find a quantitative model which produces just the right quark-antiquark imbalance. At energies of 10^{14} GeV the quarks are still free. The quark dynamics of such systems as charmonium suggest that the quarks themselves have masses of only the order of a few hundred MeV. At those high temperatures, they are effectively massless; and if a quark-antiquark pair annihilates, it is readily reconstituted. Detailed calculations indicate that some 10^{-6} second later, when the temperature has dropped to the equivalent of a GeV or so, a phase transition in which the quarks are incorporated into the hadrons once and for all takes place. There is a slight difference in number between hadrons and antihadrons. It reflects the quark-antiquark imbalance, which means that hadrons and antihadrons cannot completely annihilate each other. What is left over are the hadrons that have escaped annihilation when the antihadrons ran out; these are the hadrons we observe. The fact that quarks and antiquarks are so nearly numerically balanced accounts for the fact that the number of hadrons is so much less than the number of photons. The mutual annihilation theory is nearly complete. If theories like that are right, they predict a residue at the present epoch in the form of baryon nonconserving decays of the proton.

It was just that idea that inspired the building of detectors like the IMB; it is just there that the beautiful SU(5) theory stuck its neck out and seemingly got it chopped off by the kind of experiments done at the IMB detector. The baryon decays in SU(5) obey an interesting selection rule that is common to several but not all of the grand unified models. In SU(5) any change in baryon number must be accompanied by an equal change in the lepton number. This means that the decay $p \rightarrow e^+ + \pi^0$ is allowed and the decay $n \rightarrow e^- + \pi^+$ is forbidden. It is the former decay that the IMB detector was built to detect. The proton lifetime predicted by the simplest version of SU(5) is about 10^{30} years, whereas the experiment, as we noted in the Prologue, now sets a limit to the lifetime of at least 3.5×10^{32} years with no decay event having so far been observed with certainty. The observation of even one such event would be of extraordinary importance for this whole set of ideas and would tie together the large and the small and the old and the new.

As if poor Mr. Tompkins had not suffered enough with the massive strings and all the rest, he is about to have his worst experience yet. The heat bath in which he has been sitting, which was at a pleasant 10^{26} K at the end of the grand unification epoch, will now suddenly suffer a temperature drop of some 26 orders of magnitude, down to something like a frosty 1 K. At the same time, the universe will have expanded by a factor of 10^{26}. The dread "inflationary epoch" will have begun. It will last only some 10^{-33} second.

Before we get into the why's and wherefore's of inflation, a historical note may be instructive. The notion that under certain circumstances the universe could inflate exponentially is almost as old as the subject of relativistic cosmology itself. Indeed, in 1917, the same year in which Einstein published his first paper on relativistic cosmology, the Dutch astronomer Willem de Sitter published a second model cosmology—de Sitter space—in which the universe expands exponentially. De Sitter noted that the expansion might have observable consequences in that light from distant objects might be red-shifted, an example of the Doppler effect. He wrote, "The lines in the spectra of very distant stars or nebula [it was not yet understood that some of these nebula were actually galaxies external to the Milky Way] must therefore be systematically displaced towards the red, giving rise to a spurious radial velocity." It is not entirely clear why de Sitter called this velocity "spurious." Perhaps he had in mind that it is the velocity with which space itself is expanding, as opposed to the velocity of motion of, say, a star within a galaxy. Such a star could be sitting perfectly still in its galaxies but, because space is expanding, find itself farther and farther away from its neighbors.

The speed of that separation need not even be less than that of light, the usual relativistic speed limit on motions of material things. As Einstein was wont to say, "Space is not a thing." Indeed, during the inflationary epoch, space was expanding faster than the speed of light. It is interesting, historically, that when, in 1929, Edwin Hubble first observed the cosmological red shift and especially the fact, also noted by de Sitter, that the red shift and hence the recession velocity was proportional to the distance that separated the galaxies, he called it the "de Sitter effect." In fact, his raw experimental plot looks so scattered that one wonders if he would have found his linear relation between recession velocity and distance of separation, called Hubble's law, if he had not been looking for it.

Although Hubble's expansion law has been confirmed, and reconfirmed, many times since its discovery, the de Sitter space itself remained, until recently, more of a mathematical curiosity than anything else. The reason was that, as the dynamics of the expansion came to be understood, it became clear that, at least in the conventional models, the history of the expansion divides itself into two parts. For something like the first 10 000 years of its history the universe was in what is known as its radiation-dominated phase. That means its energy density, the average amount of energy per unit volume, is given essentially entirely by the energy of massless particles, particles whose mass is small compared to the ambient temperature of the universe.

It can be shown that this is what leads to the $R \sim \sqrt{t}$ expansion law. When the temperature drops to about 10^4 K, the contribution from the massless particles, which, it can be shown, is decreasing like T^4, is overtaken by the contribution of the massive baryons. Even though the baryons are fewer in number than the photons by a factor of at least a billion, the large rest mass energy of, say, the proton, makes up for it when the temperature falls below 10^4 K. At that point the universe becomes what is called matter-dominated. As can be shown, this implies that for most of the universe's existence after 10 000 years, the expansion law becomes $R \sim \tau^{2/3}$. Then the universe expanded somewhat more rapidly than during the radiation regime. At no time, according to this scenario, was the expansion exponential. For that to be so, the universe would have to have undergone an epoch of constant negative pressure. Until 1980 there did not seem to be either a mechanism or a reason for the universe to have undergone such a phase. Then came the paper of Alan Guth of M.I.T. Guth's paper provided the motive and suggested the mechanism.

First the motive. While experimental evidence such as the discovery of the 3 K relic radiation appears to confirm the Big Bang cosmology, it had been recognized by the time of Guth's paper that there were loose ends in the scenario that might well sink it. We shall give two examples to illustrate the genre. The first is the matter of the magnetic monopoles. As we have seen, in the grand unified scenario they will be created in profusion at the symmetry-breaking energy, of some 10^{14} GeV, and they will have masses of the same order of magnitude. Hence, unless they can be gotten rid of somehow, they will dominate the energy density of the universe and make it highly curved, in contradiction to experiment. Apart from that, despite serious attempts to observe them in the laboratory, no clearcut examples of monopoles have been found. Calculations show that monopole-antimonopole annihilation in the early universe does not operate efficiently enough to get rid of the monopoles, so we are left with the problem of finding some other mechanism.

The second example of trouble with the Big Bang scenario is quite different but equally paradoxical. It has to do with the relic microwave radiation. To say that the microwave background radiation has a temperature of some 3 K is to mean that the energy distribution of these microwave photons is identical with what would be found in a cavity whose walls had been cooled down to 3 K. The walls emit radiation which gets trapped in the cavity, and that radiation has a characteristic energy or, equivalently, a frequency distribution that depends on the temperature of the walls of the cavity but not on what the cavity walls are made of. That is the so-called blackbody spectrum, and it was to explain its shape that Planck invented the quantum theory in the first place. Figure 7-2 is a sketch of such a blackbody curve.

Since the discovery of the cosmic microwave radiation in 1965 by Penzias and Wilson—corresponding to finding one point on the blackbody curve—other experimenters have managed to fill in the rest of the curve. Moreover, similar curves have been plotted by examining different regions of the sky. If relic microwave quanta are gathered from different regions of the sky by pointing radio telescopes in different directions, the same blackbody spectrum, apart from one interesting effect, is found. There is a Doppler shift that appears to indicate that our entire galaxy is moving with a speed of about 640 kilometers per second in a direction whose significance is obscure. Once that effect is taken out, the curves taken from different sky locations agree to about 1 part in 10 000, something that is known as the isotropy of the microwave radiation. But for the various regions to produce such nearly identical microwave distributions must mean that the photons from the regions were in "communication" with each other at some time in the past; otherwise, how would they all "know" to be in blackbody distributions at the same temperature? No signal can travel between these re-

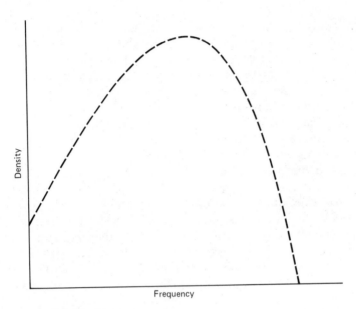

Figure 7-2. The blackbody spectrum.

gions any faster than the speed of light, and, at least in the conventional scenario, these regions were so widely separated that intercommunication would not be possible even if the light signals had traveled between them for the entire lifetime of the universe.

These two problems, and others, with the conventional Big Bang scenario seemed to be intractable. But, as Guth noted, all could be dealt with very handsomely if the universe went through a very brief phase of rapid inflation. The monopole problem, for example, disappears because the monopoles were created prior to inflation. After inflation, they found that their density in space had been reduced by a factor of something like 10^{78} simply because the volume scale of space had increased by at least that factor. They suddenly found themselves a rarity in an inflated universe. The problem of the causal connections among the different microwave regions is solved if we suppose that our present universe evolved from a single tiny causally connected region. After inflation, the universe was still causally connected and could continue to evolve to its present size. Immediately after inflation the presently observed universe would occupy a region some 10 centimeters in diameter contained in an inflated region of 10^{26} centimeters.

Guth also provided the essential mechanism which could give rise to such an inflationary phase, although getting all the details right has engaged, and is still engaging, large numbers of cosmologists. He noted that at the phase transition energy—about 10^{14} GeV—at least some of the Higgs fields would be acquiring large average values in the vacuum. Those vacuum values contribute to the energy density of the universe and indeed, at least for a brief instant, dominate it. In their cosmological effects they resemble an extra term that Einstein introduced into his original cosmological equations when he felt, erroneously it turned out, that he had to keep the universe from either expanding or contracting. This additional "cosmological constant" term acts like a negative pressure. For a brief instant, the cosmologists inform us, it becomes the dominant term in the energy and the universe inflates.

A puzzle which is still outstanding is why this term is as miniscule as it is at present. After the inflation stops, the universe, making use of the potential energy stored in the vacuum, reheats and the conventional expansion resumes. For the next 10^{-11} second nothing much happens, and Mr. Tompkins can coast along with the expansion. By 10^{-11} second, the ambient temperature has dropped to a comfortable 10^{15} K, corresponding to an energy of about a hundred GeV. According to present ideas, it is at this energy that the Higgs mesons responsible for the symmetry breaking of the electroweak SU(2) × U(1) gauge groups acquire their nonvanishing average vacuum values. When that happens, the W's and the Z^0 acquire their masses, as do the charged leptons. The neutrinos, if they have masses, may have acquired them at the end of the grand unified epoch. The coupling constants g and g' have now acquired essentially their presently observed values. Quarks and gluons remain massless for another microsecond until the temperature drops to the equivalent of about a GeV. At that point hadronic matter takes on its familiar form with the quarks and gluons now trapped inside the hadrons. Unstable mesons, like the pions and K-mesons, abound, but in short order they either decay into stable particles or annihilate with their antiparticles. The density of the amalgam is about 10^{17} grams per cubic centimeter, about 10 000 times denser than a uranium nucleus or the center of a neutron star.

During the next second there is a good deal of activity but, as far as we know, it does not involve any *outré* physics such as phase changes. By this time the temperature has dropped to an almost frigid 10^{10} K, equivalent to an MeV of energy. The heavy leptons have largely disappeared by decay or annihilation. Each time a species, like the muons, annihilate, energy is dumped into the medium because the photons that are produced in the annihilation carry the rest-mass energy of the disappearing particles. Any species that is in contact with these heated photons has its temperature raised as well. However, some species are not in direct contact with the photons and do not share this heating. The gravitons, for example, never share the annihilation heating, so that if we were ever able to detect freely expanding relic gravitons, we would find them to be colder than the relic photons.

Neutrinos present a more complex picture. As electrically neutral particles, neutrinos interact with photons only with great difficulty. But they do interact weakly with, for example, electrons, and the electrons, in turn, interact with the photons. This chain of interactions, so long as it is not broken, ensures that the neutrinos share whatever heating the photons suffer during annihilations of the charged particles.

At about this time, however, corresponding to an energy of 1 MeV or so, the chain of interactions is broken. The reason is that the rates of the neutrino-electron interactions, which depend sensitively on the temperature of the medium, fall off much more rapidly than the rate of the expansion of the universe. The reaction rates cannot keep up with the expansion, and the neutrinos uncouple from the electrons, as well as the neutrons and protons, with which they have been interacting. At about 0.1 second, when the ambient temperature has dropped below 0.51 MeV, which corresponds to the mass of the electron, electrons and positrons annihilate. Essentially all of the positrons disappear, leaving behind only enough electrons to balance the residual proton charge to preserve the charge neutrality of the universe as a whole. After the annihilation there remain about a billionth the number of electrons and protons as there are photons. Meanwhile, the neutrinos, which do not share the photon heating, continue to expand freely. If some ingenious experimenter could detect them now, he or she would probably find a number comparable to that of the photons but at a reduced temperature.

Getting back to what has been going on at 1 second after the Big Bang, a series of events first envisioned by Gamow and his collaborators begins to unfold. Mr. Tompkins can witness the working out of the theoretical predictions of his creator. That aside, we shall see in this regime a remarkable convergence of the large and the small, of cosmology and elementary-particle physics. The issue is the cosmological production of light nuclei, especially helium. The largest component in the visible universe by weight at the present time appears to be hydrogen. It may well be that some 90 percent of the mass of the universe is in "dark matter," the matter which does not give off luminous radiation. The nature of this matter is still obscure, and we shall discuss some of the possibilities shortly. Hydrogen is followed by helium, which constitutes some 25 percent by weight of the visible universe. There are also lesser amounts of other light elements such as deuterium and lithium. Helium is made all the time in stars in the fusion processes that provide the energy that keeps the stars shining.

But stellar helium production cannot account for the presence of so much of the stuff so widely distributed about the cosmos. That was clearly recognized by Gamow and his collaborators, and in the series of papers discussed at the beginning of the preceding chapter they set out to compute the observed abundances from something like first principles. They understood that the key ingredient in helium production was the ratio of neutrons to protons at any given time. The reason is that, once the temperature drops enough that the nuclei that have been formed are not torn asunder by the ambient radiation, all available neutrons will be incorporated into the light nuclei beginning, as was mentioned in the preceding chapter, with the radiative capture of neutrons by protons to form deuterium. Deuterium is just a way station in the process of making helium, since the very fast nuclear and electromagnetic reactions convert the deuterons almost at once into ^{4}He, the most stable isotope of helium which has a binding energy of some 28.3 MeV. Thus, conceptually, we can forget about the intermediate processes and assume, as a very good first approximation, that all available neutrons, at a suitable low temperature, become instantaneously converted into helium. That is why the neutron-to-proton ratio is absolutely crucial.

We can follow the evolution of that ratio from a microsecond, when neutrons and protons are first assembled out of quarks, to something like 168 seconds — roughly 3 minutes — when the ambient temperature is low enough for helium production to take place successfully. Without special pleading, there is no reason to think that the initial value of the neutron-to-proton ratio is anything different than unity. Indeed, should it have differed from unity, collisions with neutrinos and electrons in this regime will rapidly restore the balance.

To get a handle on the situation as it evolves in time, let us begin by imagining a world in which the weak interactions are switched off. In that hypothetical world, nothing intervenes to change the initial neutron-to-proton ratio, which means that at 3 minutes it is still 1. Hence in this hypothetical world all the protons find neutron mates and the fraction by weight of helium produced also is unity. There is essentially nothing but helium with a trace of other light elements.

Now suppose we switch on the weak interactions. The first effect is to make the neutrons decay, which depletes the number of neutrons and lowers the ratio. The lifetime of the neutron is about 896 seconds. If the decay were the only effect, it would lower the helium production by a factor of

$$e^{-168/896} \simeq 0.8$$

that is, by the exponential of the ratio of the time at which the helium formation takes place to the lifetime of the neutron. That is in the right direction toward getting agreement with the observed answer of about

0.25, but something else must be going on. Principally, what is going on are inelastic scattering processes of the form

$$v_e + n \leftrightarrows p + e^-$$

Since these can proceed in either direction, they would not affect the neutron-to-proton ratio if it weren't for the fact that the neutron is heavier than the proton. As the temperature drops, it becomes more and more difficult for the electron capture reactions $e^- + p \to v_e + n$ to replenish the neutrons. If this reaction proceeded indefinitely as the temperature dropped, all the neutrons would be depleted by means of the inelastic processes. But, as we have already mentioned, these reactions are a rapidly varying function of the temperature, and, at an MeV or so, they fall behind the expansion of the universe and are essentially cut off. If there were no additional effect due to the neutron instability, the neutron-to-proton ratio at the cutoff temperature would be the ratio that would be operative at the time of helium production. The ratio would be frozen at its value at the cutoff temperature. If that value is corrected by allowing the neutron to decay, a detailed calculation shows that the fraction of neutrons available at the time of helium production is about 0.126. Since every helium nucleus contains two neutrons, and since essentially every free neutron has become part of a helium nucleus, the ratio by mass of helium to neutrons and protons is twice that, which leads to the observed 0.25.

It will be noticed that the inelastic scattering reactions responsible for changing neutrons into protons in this regime involve only the electron neutrinos. The energies are too low for the muon and tauon neutrinos to play any part in this conversion. Reactions involving them would lead to the creation of massive muons and tauons, for example, $v_\mu + p \to n + \mu^-$, and there is not enough energy for that to happen. We may then ask, do the nonelectron neutrinos have any part to play in the helium production at all? The answer is yes, and that will lead us to a deep connection with laboratory elementary-particle physics. The rate of expansion of the universe is determined by its mass-energy density at any moment in time. That is plausible because the expansion is governed by gravitation. The greater the mass-energy density the faster the expansion. During the radiation regime, the mass-energy density is dominated by the contributions of the massless or nearly massless particles.

By "nearly massless" is meant that the masses are small compared to the energy equivalent of the ambient temperature, which in the regime we are discussing is at least a few MeV. The more species of such particles there are, the greater is the mass-energy density, and that is just how the other flavors of neutrinos get into the act. Each additional flavor causes the expansion to speed up. But that affects the neutron-to-proton ratio. There are two effects, and both of them *increase* the number of neutrons available for helium production. The first effect, which is less important, is entirely straightforward to describe. The faster the expansion of the universe the less time the neutrons have to decay before they enter into helium production. Hence, this effect evidently increases the neutron-to-proton ratio.

The second effect, which is more important, is a little more subtle. As we have mentioned, there comes a temperature at which the inelastic neutrino-scattering rates can no longer keep up with the expansion rate and the neutrinos uncouple from the rest of the plasma. We may call this temperature the freezing temperature because, at it, apart from the effect of the neutron decay, the neutron-to-proton ratio is essentially frozen in. Because determining the freezing temperature depends on the expansion rate and because the expansion rate depends on the number of neutrino flavors, it is not surprising that the freezing temperature depends on the number of flavors as well. It turns out that the freezing temperature increases as the sixth root of the number of flavors, $T_F \sim N^{1/6}$.

That is hardly an obvious result, but there it is. But any increase in the freezing temperature also increases the neutron-to-proton ratio. To see that, just imagine a hypothetical limiting case in which so many new species are added that the freezing temperature is nearly as high as the temperature at which neutrons and protons are created out of quarks in the first place. At that temperature the neutron-to-proton ratio is 1, which is as high as it ever gets. Thus the consequence of adding flavors is to increase the helium production, since it increases the neutron-to-proton ratio. A great deal of careful experimental and theoretical work on cosmological helium has been done to learn how many additional flavors there can be over the three we know about—electron, muon, and tauon—in light of the measured helium abundance. The answer appears to be at most *one*, and that additional flavor can be accommodated only with some strain. If all of that is correct, cosmology may be telling us that we have seen about as many flavors of neutrinos as we are going to see. Why that number should be three, or thereabouts, is one of the great unsolved mysteries in elementary-particle physics. Now to the connection with the laboratory experiments.

In the preceding chapter we described how the Z^0 was discovered. Among its several decay modes are decays into neutrino-antineutrino pairs. Indeed, the Z^0 decays into all possible flavors of neutrinos and antineutrinos. Hence, if we could make a precise measurement of the portion of the Z^0's width due to neutrino decays, we would have a laboratory determination of the number of flavors. The full measured de-

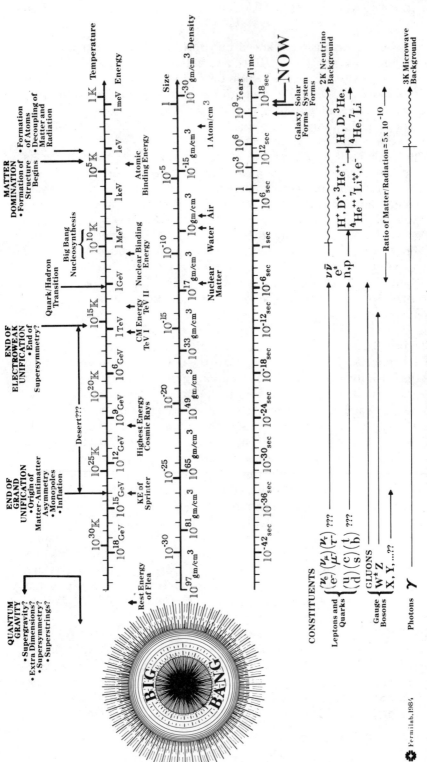

Figure 7-3. The relationships of temperature, energy, size, density, and time that existed in the very early universe; they can be traced by studying the point at which the original "soup" converted into such matter as quarks and leptons. *(A Fermilab Photo; used by permission.)*

cay width—including all the decay modes—is a few GeV, and each added flavor of neutrino contributes only about 0.15 MeV to the width. To see the effect of adding a single new species would require measurements on thousands of Z^0s, and they have not yet been made. Some work in this direction has already been done, however. The UA1 group at CERN has enough events to put a bound on the total number of flavors of about 10, meaning that, using this somewhat weak experimental bound, seven additional flavors might be possible. But it turns out that if we use the ratio of the measured W and Z^0 widths, and some theory, we can do much better. Using these semitheoretical bounds, we might tolerate only two additional flavors. Those results will certainly get more accurate with time. Even as they stand, they provide an extraordinary convergence between the ideas of elementary-particle physics and cosmology.

We have now reached 3 minutes in our history of the universe. Mr. Tompkins is beginning to see the familiar nuclei, at least the light ones, but no atoms. To have proper atoms, electrons must stick to the atomic nuclei, and it is still too hot for that. But after 100 000 years the temperature drops to about 3000 K, the energy equivalent of about one-tenth of an electron volt. Now the electrons do stick to the light nuclei. Once that happens, they can no longer scatter from the photons. Hence, the photons are released and can expand, freely cooling as they expand. After about a billion years the gravitational ripples—the places in the plasma where gravitating masses clump, perhaps, around the primordial stringy loops—begin to coalesce into galaxies. Stars are born, and the first supernovas begin to explode and spew the heavy elements into space. At about 10 billion years after that, Mr. Tompkins can be found waking up and telling his wife, Maud, that he has just had a most astounding dream. The beautiful graphic in Figure 7-3, prepared at the Fermi National Laboratory, illustrates the whole history of Mr. Tompkins' dream.

8
The Future

In 1936 Einstein published, in the *Journal of the Franklin Institute*, a beautiful and profound essay on his conception of how deep science proceeds. It was called "Physics and Reality." His main concern was to show that, on the deepest level, physics — that was his chosen example — does not proceed by simple induction. It proceeds deductively from axioms that, with constraints, can be chosen freely. As he puts it,

> The essential thing is the aim to represent the multitude of concepts and theorems, close to experience, as theorems logically deduced and belonging to a basis, as narrow as possible, of fundamental concepts and fundamental relations which themselves can be chosen freely (axioms). The liberty of choice, however, is of a special kind; it is not in any way similar to the liberty of a writer of fiction. Rather, it is similar to that of a man engaged in solving a well-designed word puzzle. He may, it is true, propose any word as the solution, but there is only *one* word which really solves the puzzle in all its forms. It is an outcome of faith that nature — as she is perceptible to our five senses — takes the character of such a well-formulated puzzle. The successes reaped up to now by science do, it is true, give a certain encouragement for this faith.

One of the "successes" that Einstein must surely have had in mind was his own general theory of relativity and gravitation, which he published in 1916. His theory takes as its basic axiom that the laws of physics should take the same form in all coordinate systems connected by general transformations of the four-dimensional continuum of space and time, hardly a self-evident proposition. Einstein then proceeded to deduce the form of the simplest theory that would be consistent with that axiom, and remarkably, that turned out to be a theory of gravitation that was not only consistent with Newton's, when the gravitational fields were weak, but also predicted entirely new phenomena, like black holes, when the gravitational fields were strong, "a certain encouragement for this faith" indeed! Einstein found that thinking of that kind produces theories that have something entirely mysterious to do with the world. In the same essay he writes,

> The very fact that the totality of our sense experiences is such that by means of thinking...it can be put in order, this fact is one that leaves us in awe, but which we shall never understand. One may say "the eternal mystery of the world is its comprehensibility."

It is easy to get carried away by this notion and many feel that, in his later career, after the discovery of general relativity, Einstein did get carried away. He seems to have come to believe that significant generalizations in physics—axioms—could be created with little or no experimental input; that pure thought would be enough. He spent the last years of his life in what seems now to have been a futile attempt to unify gravitation and electromagnetism, on a purely classical basis, by using the sorts of symmetry arguments that had led him to the general theory of relativity. Be that as it may, he brought into physics the idea, still with us, that although experience can suggest the axiomatic basis of our theories—it is a necessary component—it is never sufficient. Some leap of the imagination is required, and it is guided by criteria of symmetry or elegance, which, in most cases, the scientists involved would have great difficulty in articulating precisely.

In the preceding chapters we have seen several examples of this process. In 1954, to take one example, Yang, guided by the experimental observation that isotopic spin appeared to be a useful global symmetry, made the conjecture that *local* isotopic spin symmetry, the freedom of choice at every point in space-time of what is a neutron and what is a proton, might be an approximately valid symmetry as well. That was hardly a self-evident proposition either. It is what led him, in collaboration with Mills, to create the Yang-Mills theory. At the time it was created, few physicists, if any, shared Yang's intuition that local non-Abelian gauge invariance was going to lead somewhere. Indeed, to most physicists, it seemed to create more problems than it solved. In retrospect, however, as we read the sequence of papers, the correct papers, ignoring the false starts—the whole development from the Yang-Mills paper to quantum chromodynamics and the electroweak theory—seems so inevitable that one might, naively, wonder why it took 20 years of intense, often very frustrating, theoretical and experimental work to complete the development. The solved problem does take on, in Einstein's image, the aspects of the solution to a "well-designed word puzzle." Now that the work has been neatly and elegantly laid out, it is almost impossible to reenter the frame of mind prior to the revelation of the solution.

In addition to the satisfaction of contemplating the completed solution of a difficult scientific problem, such a reading of history also contains lessons about trying to predict the future. The simple and blunt lesson is that it is impossible. It is instructive to read conference reports and review articles of, say, the period around 1954, to stay with the Yang-Mills example for a moment, to see what clues to the future they contained. None can be found. The urgent concerns expressed in those conferences seems to us to be, by and large, antique and strange. One thinks, with some sympathy, of the countless hours of almost wasted hard work that the attendees of those conferences were about to spend following false trails and wonders how what we are doing now will appear to comparable observers 30 years hence.

The futility of trying to predict the future in science is also made evident by a contemplation of the recent history of the accelerators. To give some examples, the Berkeley Bevatron was designed to find the antiproton, which it did. But it also was used to discover a new field of strange particle resonances that no one had predicted and which, ultimately, became part of the data that led to SU(3). The Stanford electron accelerator was originally designed with the idea of continuing the elastic electron-nuclei-scattering experiments to higher energies. That was done, but the real importance of the accelerator was revealed when inelastic scattering produced convincing evidence for constitutent quarks and later, when new narrow mesonic resonances were found by colliding electrons and positrons, new flavors of quarks. The magnificent arrangement of accelerators at CERN was designed to discover the charged W mesons and the Z^0. That it did, but, in the case of the Z^0, it went on to reveal a deep connection between the number of neutrino flavors found in the laboratory and the number involved in the production of primordial helium.

Although these examples show just how difficult it is to anticipate the far future in science, there is, nonetheless, some value in discussing some of the questions we think may be answered in the not too distant future. Here we are on firmer ground, because such questions grow out of the science we know. Even if we are surprised by the science we don't know, the answers to less futuristic questions will still be important. Let us begin with what is perhaps the outstanding puzzle in cosmology, especially since it is a puzzle that may well have deep ramifications for the physics of elementary particles. That is the puzzle of the "dark matter," the "missing light," a puzzle that has both an experimental and a theoretical character.

Let us begin with the theory. As we have mentioned, observations show that, at the present time, the universe is very nearly flat. Another way to state that fact, in terms of the expansion of the universe, may be a little less abstruse. The cosmological equations of motion that govern the expansion, at least the ones that are commonly used, admit three kinds of expansions. There are expansions corresponding to negative curvature that will slow down and then stop, in a finite time, and lead to a recollapse of the universe, perhaps into its original singular state. There are expansions corresponding to positive curvature that will continue forever, and there is the zero curvature expansion, the dividing line between the first two, which will continue forever, but with a speed of recession that approaches zero. The present speed of recession is caled the Hubble constant. That is something of a misnomer, because it is not a constant; it differs from epoch to epoch. If the curvature is zero, it turns out that the square of the Hubble constant is proportional to the mass-energy density of the universe. If the curvature is not zero, there is an additional term that involves the curvature as well.

To measure the Hubble constant, it is necessary to measure the velocity of recession of distant galaxies. That is the rub, since there is considerable uncertainty among astronomers about those distances. The measured Hubble constant is uncertain by a factor of something like 2. If we knew it, we could predict the mass-energy density needed for a zero-curvature expansion, a density that is usually called the critical density ρ_c. If h represents the uncertainty in the Hubble constant, then, using the known numbers,

$$\rho_c = 1.88h^2 \times 10^{-29} \frac{\text{grams}}{\text{centimeter}^3}$$

where $1 \geq h \geq 0.4$ represents the experimental uncertainty in h. The quantity that concerns cosmologists is the ratio of the observed mass energy density ρ to the critical density ρ_c. The ratio ρ/ρ_c is a quantity that is called Ω. If $\Omega = 1$, then the universe is flat. That, perhaps, is the most theoretically elegant value of Ω, but then we must check experimentally to see if in reality Ω does equal 1.

The problem is that what you see is pretty much what you get, and there may be a lot that you don't get. The density of *visible* matter is only a few tenths of the critical density at most. That makes the problem still more tantalizing, because it is difficult to understand why the density should, at the present time, be so close to the critical density if it is not exactly at the critical density. And that would lead one to conjecture that there should be dark matter. Experimentally, however, astronomers have known for some time that there *is* dark matter. Galaxies are known to rotate around their nuclei—their galactic centers—which probably reflects the angular momentum of the gas cloud out of which they were originally created. This kind of rotating collection of stars should obey a simple law of gravitation that relates the distance of the star from the center of the galaxy to its orbital velocity. According to that law, the velocity of these stars should fall off inversely as the square root of the distance from the galactic center.

In fact, when the measurements are carried out on the odd visible galactic star that lies beyond the bulk of the visible matter, there is no evidence at all that the inverse-square law is obeyed. In all known cases, the velocities remain constant as the distance increases. That appears to mean that these stars are in the midst of "dark matter" parts of their galaxies. In fact, most of the mass does not seem to be visible. But what is it? Knowing the answer would perhaps resolve one of the most vexing enigmas in present-day cosmology, that is, the value of Ω.

Among the candidates for the dark matter, one of the most appealing has been the neutrino. In the first place, the cosmological origin and subsequent history of the neutrinos persuade us that there should, at present, be about as many of them per cubic centimeter, some 400, as the relic photons. They are, because of their weak interactions, impossible to detect by present-day methods, although futuristic ideas have been proposed. Finding them would be of significance comparable to that of the discovery of the microwave relic radiation. If they are massless, they would not help in the dark matter problem; but if they had masses as little as 10 eV, they would solve the problem.

We are much more likely to have definitive experimental results on the masses of terrestrially produced neutrinos in the near future. We do know that these masses, adding up all flavors, cannot exceed about 100 eV because if they did, they would be in conflict with the observed near flatness of the universe. The neutrino which is most accessible to mass studies is the electron neutrino.

As we mentioned in the first chapter, Fermi, in his very first paper on the neutrino in β-decay, anticipated the possibility of detecting the neutrino's hypothetical mass indirectly by looking for changes in the shape of the β-decay electron spectrum. The high-energy end of the β-spectrum is especially sensitive to such changes because those particular decays correspond to the emission of the lowest-energy neutrinos, decays whose kinetic energies are most comparable to their putative rest masses. The most attractive nucleus to use for these experiments is tritium—superheavy hydrogen—which consists of two neutrons and a proton. It decays into a light isotope of helium ^3He with the emission of an electron antineutrino; the reaction is $^3H \rightarrow {}^3He + \bar{\nu}_e + e^-$. That is good nucleus to use because the decay is not too energetic—it has a maximum electron energy of 18.6 keV—and because tritium is a relatively simple nucleus whose quantum mechanics can be computed reasonably well.

In 1980 a group of experimenters working in Moscow actually reported finding an antineutrino mass of at least 14 eV by studying the tritium β-decay. That set off a great flurry of experimental and theoretical activity, but no one has been able to confirm the Russian result so far. The best that can be said from these experiments, at the moment, is that if the electron antineutrino does have a mass, its mass cannot be larger than about 20 eV. In the meanwhile, the neutrinos that arrived from the supernova on February 23, 1987 presented an entirely novel way to try to measure the mass.

To understand the essence of the idea, suppose that all the neutrinos from the supernova were emitted at the same instant. That is not really true, but it will enable us to grasp the point. Furthermore, suppose that the neutrinos had zero mass. Then, no matter what their energy, they would all move at precisely the speed of light, because all massless particles move at the speed of light. Hence, according to this hypothetical scenario, all the neutrinos would arrive here at precisely the same instant. Next let us ask how this scenario would be altered if we gave the neutrinos a mass. If they all had the same energy, the scenario wouldn't be altered. All the neutrinos would move with the same velocity, albeit one that is less than that of light, and they would all arrive at the IMB detector, or wherever, at the same time.

But, in fact, the neutrinos that arrived at the IMB detector did *not* have the same energies; they differed in energy by as much as a factor of 2. If they are massive, the low-energy ones move slower than the high-energy ones; and even if all of them had begun their voyage at the same time, they would have arrived here at different times. We know that the eight neutrinos that arrived at the IMB detector from the supernova did so in less than 6 seconds after they had been traveling for 163 000 years. That fact alone tells us that their rest mass must be very small, because the time spread was only 6 seconds despite the fact that the energies differed by a factor of 2 from one neutrino to another.

To make all this quantitative is more difficult, because one must take into account that the neutrinos are not emitted at precisely the same time. The time spread in the emission can be estimated theoretically only by using a model of the supernova explosion. It is a pity that there were so few events—eleven from the Kamiokande detector in Japan and eight from the IMB. With all the uncertainties, these results nonetheless suggest that a mass of about 20 eV would be acceptable. But a zero mass neutrino also is compatible with the results if the emission of the neutrinos was suitably spaced out. We cannot predict when the next supernova explosion will occur in or near our galaxy, but in view of all the experiments now underway to measure the neutrino mass from β-decay we can hope that the limits we have now can be substantially improved in the not too distant future.

In the same vein, we are entitled to hope that the matter of the neutrino "oscillations" also will be settled before long. To understand this intriguing prospect, we may remind ourselves of the discussion in Chapter 3 of the K^0-$\overline{K}^0$ system. It will be recalled that the K^0, when it is produced in a strong interaction, is initially in a state of definite strangeness, $+1$ in this case. But since the weak interactions do not conserve strangeness, it can mix, through weak interactions, with its antiparticle the $\overline{K}^0$, which has a strangeness -1. Out of this mixing two new states, K_L and K_S, result. They do not have a definite strangeness, but they do have definite masses. The two states have different decay modes. In the course of time, if a beam of K^0 particles is followed by stationing observers at different points along the beam, the decay products can be watched shifting back and forth from one set of decay modes to the other. That is the phenomenon of oscillations. In fact, it is by detecting oscillations that the incredibly small mass difference of 10^{-5} eV between the two states is detected. For there to be oscillations at all, the two states K_L and K_S must have different masses.

Now a similar phenomenon would be expected to occur with massive neutrinos, except that what is involved here is, not strangeness, but flavor-changing interactions. These interactions mix neutrino flavors. It should be noted, however, that neutrinos of different flavors might in principle have different masses but still not intermix. However, if the origin of these masses is through a Higgs mechanism, then such mixing is a very natural thing. If there are both mass differences and mixing interactions, then, for example, an electron neutrino produced in β-decay could oscillate into a mu-neutrino. The amount of oscillation generated would depend on the hypothetical mass difference. Actually, it would depend on the difference between the squares of the masses of the two neutrinos in this example and on a parameter, the so-called mixing angle, that could be computed if a detailed theory of the mixing mechanism were available. The conditions of the experiment can be adjusted to observe mass differences of varying degrees of smallness. The further we move from the source the more opportunity there is for the oscillations to develop.

To cite a pertinent example, a nuclear reactor produces electron antineutrinos with typical energies of an MeV or so. If the difference of the square of the masses were, say, 10^{-2} eV, a detector would have to be placed about a football field's distance away to benefit from the oscillations. The way such a detector might work is to have it detect the flux of electron antineutrinos arriving at the source by having them produce positrons in reactions like $\bar{\nu}_e + p \rightarrow e^+ + n$. If there were oscillations, would find that some of the electron antineutrinos had "disappeared" on their way to the source, meaning that they had oscillated into muon antineutrinos which cannot produce this inverse β-decay reaction. For more energetic neutrinos coming from accelerators, we would try to see, for example, if muon neutrinos from a muon neutrino beam, after oscillations, could make electrons. Experiments of both varieties have been made, and despite flurries of enticing rumors, no such effect has as yet been definitely observed. This matter, hopefully, will be resolved in the not too distant future.

In this regard, and others, we would be remiss in not discussing the neutrinos from the sun. The sun functions as a giant fusion reactor in which light nuclei fuse together to make somewhat more massive nuclei with the release of energy. There is a *release* of energy because, when it comes to nuclei, the whole nucleus is less massive than the sum of its parts, the difference being the energy that binds the nucleus together. It would serve no point here to run through all the reactions by which energy is produced in the sun. Instead, we list two important reactions in which *neutrinos* are produced, along with the average and maximum energies of the emitted neutrinos. The first reaction produces about 10 000 times more flux of neutrinos than the second. It is the fusion of two protons into a deuteron and leptons:

Reaction	Average energy, MeV	Maximum energy, MeV
$p + p \rightarrow D + e^+ + \nu_e$	0.26	0.42

Because of their relatively low energies, these neutrinos are very difficult to detect in terrestrial experiments. We shall come back to future prospects for their detection a bit later after we discuss the second reaction, which does seem to have been detected. It involves the β-decay of an isotope of boron into an isotope of beryllium:

Reaction	Average energy, MeV	Maximum energy, MeV
$^8B \rightarrow {}^8Be + e^+ + \nu_e$	7.2	14

These neutrinos, although substantially less numerous, are much more energetic, which increases their prospects for terrestrial detection. The trick is to use an inverse β-decay reaction that is readily triggered by neutrinos in this energy range. The Soviet physicist Bruno Pontecorvo first suggested such a reaction in 1946. The reaction he proposed involves the absorption of these solar neutrinos by an isotope of chlorine to produce an isotope of argon:

$$\nu_e + {}^{37}Cl \rightarrow {}^{37}Ar + e^-$$

This is a particularly fortuitous choice, because large amounts of chlorine can be assembled in the form of the common cleaning fluid CCl_4, carbon tetrachloride. Ray Davis, formerly of Brookhaven and now of the University of Pennsylvania, and his collaborators have been using this substance in a giant tank located 4850 feet below ground in an abandoned gold mine in Lead, South Dakota, the Homestake Mine. His reasons for going deep underground are the same as those of the people at the IMB, namely, to avoid cosmic-ray background contamination. His experiment makes use of some 600 tons of cleaning fluid. Typically, he exposes the tank to the solar neutrinos for about a month at a time. At the end of the month he extracts whatever ${}^{37}Ar$ has been formed in the inverse β-decay, something he can do with a very high efficiency. The result—the number of solar neutrino events—is given in terms of the solar neutrino unit (SNU).

It is not important for us to go into the details of what the SNU is. What *is* important is to note that theoretical estimates of the unit, on which a huge amount of work has been done, give a prediction of somewhere between 5 and 7 SNU, whereas the observed rate is only about 2 SNU. The discrepancy has been known for some time, and it has given rise to another cottage industry of intense theoretical work. Some theorists have argued that the solar model that is being used is slightly wrong. It turns out that the number of predicted SNU is an extremely sensitive function of the central temperature of the sun, and it might be that we do not know that temperature well enough.

Another idea, however, is the neutrino oscillations. All the neutrinos being produced in the sun are initially electron neutrinos. They must propagate through the solar interior and then into space in order to get to Lead, South Dakota. Propagation through matter can, under certain circumstances, very much enhance the oscillation effects. The enhancement effect is so pronounced, in certain cases, that the difference between the squares of the masses could be less than 10^{-8} eV^2. If the three species were oscillating, it would be possible to reduce the initial electron neutrino flux by a factor of 3, enough to make the theory and the experiments agree. Such a small neutrino mass would be useless as far as its contribution to the dark matter problem is concerned, but it is vitally important as a matter of principle.

In the not too distant future, Davis's experiments will be complemented by others using different detectors. As a matter of fact, one reason for the Kamiokande detector was to detect solar neutrinos, because the Kamiokande has a lower energy threshold than the IMB detector has and, in the near future, it should function in that mode. Gallium seems to be a very promising material for a solar neutrino detector. It should be sensitive to the lower-energy, higher-flux neutrinos produced in the pp reaction given above. Large gallium detectors are now being built in various facilities. We should know in the not too far distant future whether we are dealing with a flaw in the solar models or something more fundamental concerning the properties of the neutrinos.

The experimental devices mentioned are passive in the sense that, once they are built, the experimenter can only wait until something happens—an external event like a neutrino arriving from the sun or a supernova—before witnessing any new physics. But elementary-particle physicists and cosmologists, an overlapping community these days, are eagerly awaiting the construction of new active experimental devices, that is, accelerators. The three most striking possibilities are the large electron project (LEP) under construction at the CERN Laboratory near Geneva, the superconducting super collider (SSC) now under active review and discussion in the United States, and the Stanford linear collider (SLC) now being readied for operation at Stanford.

The LEP, which is funded by the same consortium of European states that funds CERN, is an electron-positron collider which will produce collisions with a total energy of about 200 GeV. It should be a factory for producing W's and the Z^0. The collider ring, which is about 17 miles in circumference and, like the rings in all these accelerators, is buried underground, has the interesting property that it is in two countries, France and Switzerland. On each circuit, the electrons and positrons will cross the Franco-Swiss border four times. As we have seen, an electron-positron collider is a good machine to use for studying the production of new leptons, if there are any. Up to this time the quark-lepton analogy has held, which is to say that each time a new flavor has been discovered, it has been discovered in both the quark and the lepton sectors. If the cosmological analyses of primordial helium production are right, since the analysis allows for at most one more neutrino flavor, we would expect at most one more quark flavor. This matter may be settled by the SLC, which will generate 100-GeV collisions of .electrons and positrons.

The SSC, as it is presently conceived, is a purely American project. By September 1987, the Department of Energy had received 43 proposals from 25 states. Obviously, the SSC was clearly regarded to be a very high prestige item for a state to acquire. Contrast the public perception of this kind of physics with that of some 40 years ago, when the *New*

Yorker published a celebrated James Thurber drawing of a bevy of be-fuddled looking, bearded, physics professors being taken prisoner by those impossibly imposing Thurber women. The present-day version might show the governor of one of the twenty-five competing states at-tempting to capture the SSC. On November 10, 1988, the Department of Energy announced that the collider will, if it is funded, be built in Texas. It will surround the town of Waxahachie.

The proposed machine is estimated to cost some $4.7 billion, not counting the extra $1 billion or so the detectors, computers, and the like are expected to cost. It can be operated in either a pp or a $p\bar{p}$ colliding mode. Each particle is expected to carry a maximum energy of 20 TeV, or 20 000 GeV, meaning that collisions can take place with a total energy of 40 000 GeV. The circumference of the proposed main ring is over 52 miles! It is called a superconducting supercollider because the magnets, all 10 000 of them, are to be wound with superconducting wires. An ideal superconductor can pass an essentially infinite amount of current with no electrical resistance. A great deal of current can be passed through realistic superconductors with very little resistance. The hitch is that, with the state-of-the-art superconducting materials, the wires must be refrigerated to the temperature of liquid helium. There are exciting new materials, which have much higher superconducting temperatures, but their development is still in its infancy. The use of superconducting magnets is essential in the SSC because, for smaller amounts of current, they can be much more powerful, and that allows for a 20-TeV machine with a circumference of "only" 52 miles.

The choice of the 20-TeV energy was not accidental. To understand something of its significance, we must backtrack to the electroweak uni-fied model we last saw in Chapter 6. As beautiful and successful as it has been in its phenomenological applications, it has an Achilles heel which might, in the end, turn out to be a fatal or at least very serious difficulty. In the original version of the model, as we saw, a neutral scalar Higgs meson which has a nonvanishing average value in the vacuum is intro-duced. This is what provokes the symmetry breaking, but it leaves over a physical, massive scalar meson, the H^0, with well-defined interactions with leptons, gauge mesons, and quarks. It can, in principle, be pro-duced in $p\bar{p}$ or e^+e^- collisions in association with hadrons. Depending on its mass, it can decay into all of the above and, if it is massive enough, it can even have exotic decay modes like $H^0 \rightarrow Z^0 + Z^0$, which should present a striking signature.

As we mentioned in the preceding chapter, there are some theoretical bounds on the Higgs mass in the standard model. The main assump-tions that go into the derivation of these bounds are the conservation of probability—something that the quantum mechanicians call unitarity—

and a more debatable assumption that the Higgs meson interactions are weak enough that they can be correctly expanded in powers of the Higgs coupling constants. As a first guess, it is natural to try such an expansion, which, as we have seen, was spectacularly successful in quantum electrodynamics. If we apply it here and it works, we can argue pretty convincingly that the Higgs mass must be less than 1 TeV. That limit does not apply to the Higgs particles that might be associated with the breaking of the grand unified symmetry if there is one. These Higgs mesons would be expected to have masses of the order of 10^{14} GeV.

Such an experimental limit on the electroweak Higgs mesons presents a very attractive prospect for the supercollider. The SSC will readily produce the mesons if they have a mass of that order. If a single neutral Higgs meson with a mass less than 1 TeV is found, it will be a potentially spectacular confirmation of the symmetry-breaking mechanism of the standard model. If several Higgses of varying masses are found, it may well give us the clue we need as to how to alter the model. If the standard model's Higgs is found to have a mass greater than 1 TeV, it means that perturbation theory has broken down, and that will send the theorists off to their drawing boards. If no Higgs particles are found, it means that some other symmetry-breaking mechanism must be at work, and that also will send the theorists back to their drawing boards.

From the point of view of learning something very significant, it is a no-lose situation. In the same vein is the matter of the quarks. If the whole picture is to hang together, we must begin to find evidence of the top quark, which, if it exists, has eluded us because of its mass. If it exists, it opens up an entirely new domain of charmonium-like spectroscopy. If it doesn't exist, the present extraordinarily successful quark-lepton analogic picture goes out the window. We have, incidentally, only scratched the surface of bottom-quark physics. There are preliminary indications that the B^0 meson, a bottom-down quark composite with a mass of 5275 MeV, mixes with its antiparticle in a manner reminiscent of $K^0 \overline{K}^0$ mixing. That opens up the prospect of an entirely new experimental terrain for violations of CP invariance. It may give us clues to the origin of this very mysterious interaction which, as we have seen, may ultimately be responsible for the preponderance of matter over antimatter.

Experiments at the supercollider, when it is built, will explore distances as small as 10^{-18} centimeter. Those distances, as incredibly small as they are, are still 15 orders of magnitude larger than the Planck length. (The Planck length is the Planck time, 0.54×10^{-43} second, multiplied by the speed of light, which is about 3×10^{10} centimeters per second.) At that distance—about 10^{-33} centimeter—we expect that gravitation will be the overwhelmingly dominant interaction. For that reason, theorists have begun facing up to the apparent fact that, within conventional field theories, there does not seem to be any way of formulating a consistent quantum theory of gravitation.

The conventional theories of gravitation have in common that the fields that describe the particles are assigned values at all points in the space-time continuum. In classical physics that causes us no grief because, on the classical level, we have no reason to doubt that we could actually carry the assignment out operationally. As classical physicists we would assume that we could, literally, measure the value of a field variable at any arbitrary point in space and time, which, in turn, we could specify with arbitrary precision by using rulers and clocks. Quantum theory has, however, taught us that such an assignment cannot be made with arbitrary precision. Nonetheless, we can work with fields given in terms of space-time points, local fields, if we are careful about what aspects of those fields we actually claim to measure. We must, for example, be careful about correlations of the fields at different space-time points, since if we take the space-time points too close together, the correlations can become ill defined. This is the other side of the coin to the problem, discussed in Chapter 2, of the momentum and energy functions in the theory blowing up when the values of the momenta and energies are taken to be arbitrarily large. It was to deal with that bad behavior that the renormalization program was developed. The quantum theories of gravitation that are based on the same assumptions of localizability cannot be renormalized at all. New and terrible infinities show up in each order of the expansion in greater and greater profusion.

One way to look at this unhappy situation is to recognize that this kind of locality may be meaningless in the presence of very strong gravitational fields. Space and time may be so distorted by strong gravitation that trying to formulate a field theory in terms of these parameters may indeed be meaningless. Some physicists have responded to this dilemma by replacing the space-time continuum by a discrete lattice of space-time points. That poses extremely complicated computational problems when an attempt to introduce something like Einstein's theory of gravitation into such a lattice is made and still more complicated problems if an attempt to quantize it is made. The jury is still out on those attempts. The jury is also still out on a different approach that has attracted an enormous following among the more mathematically inclined of the elementary-particle theorists. Indeed, just as the invention of quantum mechanics kept generations of pure mathematicians busy trying to explain rigorously what the physicists had been doing with their home-spun intuitive methods, so, at the very least, this latest work also will keep mathematicians busy for decades.

The enterprise goes under the rubric "string theory," since the elementary entities in these theories are not point particles—particles whose properties are specified at every point in the space-time continuum—but are instead one-dimensional objects: strings. As far as anyone knows, these strings have nothing to do with the cosmic strings generated at the cosmological phase transitions. They do not represent individual particles; instead, they represent an entire spectrum of particles. An image that is sometimes given is that of a violin string, which represents, not a single note, but rather a single note and all its harmonics. The difference is that these new strings are meant to have dimensions of some 10^{-33} centimeter. The stunning discovery, made a few years ago, is that the theory of closed strings—string loops—resembles Einstein's theory of gravitation at large distances and is renormalizable.

Giving the primary objects extension apparently tames the infinities. This success encouraged the theorists who work in this domain to try to combine string theory with another speculative idea, so-called supersymmetry. The idea goes back several years. It is the notion that our elementary-particle theories appear to have associated with them two entirely distinct types of symmetry. There are the symmetries of space and time—the continuous symmetries of the theory of relativity—and there are the so-called internal symmetries. The latter, like isotopic spin, are symmetries which involve exchanging labels among various particles. It seemed unnatural to the theorists concerned to have two such apparently disjointed kinds of symmetry in a single theory, so they tried to put them together by creating a larger algebra that involves both the space-time symmetries and the charges that generate the internal symmetries.

Like many somewhat eccentric marriages, that one produced eccentric offspring. Each of the familiar standard model particles has a "super partner." If the standard model particle is a boson, for example, its *Doppelgänger* is a fermion. That has given rise to a plethora of new names which test one's tolerance for whimsy. To give a flavor of the genre, we have, as partners to the W's, "winos," and to the Goldstone, the Goldstino, to the Higgs the Higgsinos, and so on. None of these particles have been observed and, indeed, there is no evidence at all that such a supersymmetry plays any role in the physics we know anything about. One of the things that the supercollider can do for us is test for the existence of this new class of objects. However, the practitioners of string theory have discovered that if to the criterion for a suitable string theory the proviso that the theory be supersymmetric, that it be a theory of superstrings, is added, the number of theoretical possibilities is limited enormously.

Such theories may exist, it turns out, only in higher dimensions. The popular choice, at the moment, is ten. The ten-dimensional theory has nine spatial dimensions and one time dimension. Our empirical world has, as far as we can tell, only three spatial dimensions and one time dimension. To understand what the string theorists have in mind to do about their extra dimensions, think of the distinction between a cylinder and a line. If the radius of the cylinder is small enough, there is no practical distinction between it and a line even though, in principle, a line is a one-dimensional figure and a cylinder has three dimensions: If the radius of the cylinder is extremely small, the cylinder appears to us to be a line.

In the same spirit, it is argued, the superfluous spatial dimensions might be restricted to sizes of the order of the Planck length. They might play a role, some speculations run, only prior to inflation. During inflation, for reasons a future theory will hopefully instruct us about, only the observed spatial dimensions inflate while the others remain embryonic. There is even some hope, these theorists assure us, that the structure of the six unused spatial dimensions contains the secret of the pattern of the flavors. It will be extremely interesting to see if, in some not too distant future, the theorists can come up with a description of a crucial experiment, or experiments, that might test these very abstract ideas.

Here then are a few inklings of how the near future of elementary particle physics might look. But, in the spirit of the sixth century B.C. Taoist master Lao-tzu, we must always remember that the future that can be written is not the real future. The real future is that which *cannot* be written.

References

Alvarez, Luis, W.: *Alvarez: Adventures of a Physicist*, New York, Basic Books, 1987.

Bernstein, J.: *Einstein*, New York, Viking, 1973.

— — — and G. Feinberg: *Cosmological Constants*, New York, Columbia University Press, 1986.

Brown, L. M., and L. Hoddeson (eds.): *The Birth of Particle Physics*, New York, Cambridge University Press, 1983.

Feinberg, G.: *What's the World Made of? Atoms, Leptons, Quarks and Other Tantalizing Particles*, New York, Doubleday, Anchor Books, 1977.

Feynman, R.: *"Surely You're Joking, Mr. Feynman,"* New York, Norton, 1985.

Gamow, G.: *The Creation of the Universe*, New York, Viking, 1952.

Pagels, H. R.: *The Cosmic Code*, New York, Simon and Schuster, 1982.

— — —: *Perfect Symmetry*, New York, Simon and Schuster, 1982.

Pais, A.: *Subtle Is the Lord*, New York, Oxford University Press, 1982.

— — —: *Inward Bound*, New York, Oxford University Press, 1986.

Sciama, D. W.: *Modern Cosmology*, Cambridge, Cambridge University Press, 1971.

Weinberg, Steven: *The First Three Minutes: A Modern View of the Origin of the Universe*, New York, Basic Books, 1977.

Wilczek, F., and B. Devine: *Longing for the Harmonies: Themes and Variations from Modern Physics*, New York, Norton, 1988.

Index

About the Author

Jeremy Bernstein received his Ph.D. from Harvard University. He has taught and lectured at universities here and abroad on the theory of elementary-particle physics and cosmology. A frequent contributor to *The New Yorker,* Bernstein is the author of numerous physics papers and twelve books, among them *Analytical Engine, Einstein,* and *Three Degrees Above Zero.*

He has also written a column for the *American Scholar,* "Out of My Mind." He has received numerous awards for his science writing, including a Westinghouse AAAS Science Writing Prize, a Brandeis Creative Arts Medal, and a National Book Award nomination for his biography of Einstein.

Bernstein is a professor at Stevens Institute of Technology in Hoboken, New Jersey, and an adjunct professor at The Rockefeller University. He lives in New York City.